James M. Russell has a philosophy degree from the University of Cambridge, a post-graduate qualification in critical theory, and has taught at the Open University in the UK. He currently works as director of a media-related business. He is the author of Brief Guides to *Philosophical Classics*, *Spiritual Classics*, *Business Classics* and *Self-Help Classics*. He lives in north London with his wife, daughter and two cats.

Also by James M. Russell:

A Brief Guide to Spiritual Classics
A Brief Guide to Philosophical Classics
A Brief Guide to Business Classics
A Brief Guide to Self-Help Classics

A Brief Guide to Smart Thinking

From Zeno's Paradoxes to Freakonomics

James M. Russell

ROBINSON

ROBINSON

First published in Great Britain in 2020 by Robinson

Copyright © James M. Russell, 2020

5 7 9 10 8 6 4

A CIP catalogue record for this book is
available from the British Library

ISBN: 978-1-47214-368-6

Typeset in Garamond by SX Composing DTP, Rayleigh, Essex
Printed and bound in Great Britain by Clays Ltd, Elcograf S.p.A.

Papers used by Robinson are from well-managed forests
and other responsible sources

Robinson
An imprint of
Little, Brown Book Group
Carmelite House
50 Victoria Embankment
London EC4Y 0DZ

An Hachette UK Company
www.hachette.co.uk

www.littlebrown.co.uk

Contents

CONTENTS

CONTENTS

Introduction

The aim of this book is to give a brief, chatty, accessible overview of seventy classic smart-thinking books. While 'smart thinking', or using the power of the human brain more effectively, is now treated as a genre in itself, it has been around in one form or another for millennia. It was a feature of early works of self-improvement and philosophy and can be found in historically interesting accounts of how the mind works.

The titles we have included go back as far as Aristotle's *Rhetoric*, Epictetus's *Enchiridion* and Bertrand Russell's charming *ABC of Relativity*, and proceed through Edward de Bono's *Lateral Thinking* and into the digital era with titles such as *The Shallows, Wired to Create, Sapiens* and *Big Data*. The titles covered include thought-provoking classics on psychology, mindfulness, rationality, the brain, mathematical and economic thought, and practical philosophy.

The books are mostly presented in chronological order, showing not only how the genre has developed and grown over time, but also throwing up interesting juxtapositions and connections along the way. The two exceptions are Bryan Magee's *The Story of Philosophy* and *The Path*, a fascinating overview of ancient Chinese philosophers, included near the start as a reminder that there were many centuries in which people were discussing and writing about the same issues as modern and contemporary writers.

The main focus of the book is on the twentieth and twenty-first centuries, and on practical titles rather than pure philosophy or speculation about how the brain works. In particular there are many titles that focus on how we overcome our natural limitations and cognitive biases and maximise our ability to use our brains effectively.

In this respect it is interesting to recount an exchange between the psychologist and writer Daniel Kahneman (see p.144) and the Harvard psychologist Steven Pinker (see p.61) from a 2014 Guardian article. Kahneman expressed his pessimistic view of our cognitive limitations: 'The idea of human nature with inherent flaws was consistent with a tragic view of the human condition, and it's a part of being human that we have to live with that tragedy.'

Pinker's response was to take a glass-half-full approach: 'We have the means to overcome some of our limitations, through education, through institutions, through enlightenment.' And in many respects this sums up the basic intention of the smart-thinking genre. A writer such as Kahneman is a genius when it comes to identifying our flaws, but we also need writers who teach us ways of overcoming or getting around those flaws.

There are a couple of notable biases in the selection of books included here. One is that there is a high proportion of American (and in general, western) writers – this is somewhat difficult to avoid as so many of the bestselling titles in the genre are in that tradition. More interesting is the bias towards male writers. The fact is that this is a genre in which relatively few female writers tend to make contributions. My wife's explanation for this is that women tend to use their common sense in getting on and doing things, whereas men sit down and make elaborate plans and feel the need to share their wisdom and advice with the world afterwards. You can make of that theory what you want . . .

What constitutes a classic? That's a tough question. When it comes to recent books, in particular, it can be hard to guess which will stand the test of time. And not all of these books are even any good. It has sometimes also seemed appropriate to warn you off some pretty weak titles. The main thing that these books have in common is that they are genuine classics, bestsellers or have been highly praised. The intention of the summaries is to give you an idea of what the book is like, and whether or not it would be useful to you. The 'Speed Read' for each book delivers a brief and occasionally irreverent sense of the main thesis. You can sit down and read this book from start to finish or feel free to dip in, explore writers, and ignore entries. With luck you will find yourself with a shortlist of books you want to read in full.

Rhetoric

Aristotle, fourth century BC

Aristotle, often known as the father of western philosophy, has a tendency to write dense texts that carefully categorise an area of knowledge. In this pursuit he employs the careful precision of a librarian filing books away.

One of his most interesting works, *Rhetoric*, also known as *Ars Rhetorica*, is a treatise on the art of persuasion. He broadly defines rhetoric as the ability in any particular case to see the available means of persuasion.

The ancient Greeks had three ways of establishing possible or certain truth: logic, which they thought could lead to certain knowledge (as in a maths proof); dialectic, the art of argument in conversation (as practised notably by Plato and Socrates); and rhetoric, which is the art of using argument in speeches.

This is principally a book that examines ways of making a persuasive argument in a speech, presentation or sales pitch. For discussions of and demonstrations of dialectic, the Socratic dialogues are a good place to start.

There are three books within the work. Book one provides an overview and carefully defines and discusses various different types of rhetorical device. Book two, the heart of the work, looks at the fundamental persuasions available to an orator, while book three focuses more on elements of style, delivery, and structure.

The three types of persuasion are ethos, pathos and logos. Logos, which has the briefest treatment, is essentially the art of making a logical argument. Aristotle gives a fairly technical but precise overview of methods – such as inductive argument and abductive argument.

Ethos is defined as persuasion rooted in the character of the speaker – we are obviously more easily persuaded by people who we see as trustworthy. Aristotle breaks this down further by discussing the main qualities that make us trust someone: good sense, good moral character, and good will. If someone comes across as having good sense, it means they seem to be rational and reasonable, and are speaking in a calm, collected way. It is also most likely that we regard someone as having good sense if they can demonstrate that they are experienced and knowledgeable about the subject they are discussing. Good moral character is rooted in how we think someone behaves, both when they are aware of our presence and when they are not. And we believe someone is showing good will if they can demonstrate they are taking our interests into account and responding to them.

Of course, someone can display all of these qualities and still fail to be persuasive – which is where pathos comes in. This is the means of persuasion that relies on passion and emotional appeals: we are more susceptible to persuasion in some moods than others. A shop tries to create an atmosphere in which the right emotions are aroused that lead a shopper to make purchases. And if you want someone to support a new regulation or law, it might help if you make them feel sorry for people who are being adversely affected by the current situation, or angry at people who are unfairly profiting.

Newspapers and the media rely especially on the latter strategy: the onslaught of 'daily hate' articles that are the mainstay of certain tabloid newspapers and TV channels are entirely rooted in the need to arouse anger in their readers and viewers.

As ever, Aristotle helpfully breaks down the emotions in a range of binary choices. Anger is in opposition to calm. If people are treated with contempt, spite or made to feel ashamed then they will tend to be angry, whereas calm is more easily inspired in the absence of these things.

Friendship is seen as being in opposition to hatred: when people seem to have our interests at heart. Fear is opposed to confidence: think of the way that insurers sell their products by inspiring fear of the worst outcome. If we are confident something won't happen or we have ways of dealing with it, we can overcome fear.

The other emotional poles discussed by Aristotle feature shame vs shamelessness (for instance, we can be made to feel ashamed about morally weak or bad things we have done); kindness as opposed to unkindness (people often interpret this choice to be whether or not we would go out of our way to help them); pity vs indignation (we feel pity if someone is suffering needlessly, and indignation when we feel someone is getting a reward they don't deserve); and envy vs emulation.

This last choice is an interesting one. The essential difference is how enabled we feel. If someone on a similar level to us gets a piece of good fortune, then we are likely to feel envious if we think we should have got that lucky break ourselves or if we feel there is no chance of us having the same good fortune. By contrast, if someone on the same level as us has a stroke of good fortune that we believe could happen to us in future, then a very different emotion is inspired. We are likely to try harder to emulate their good luck.

The material on pathos is probably of the most interest to contemporary readers. As humans we are clearly very emotional creatures. The decisions we make when we are angry or fearful will be very different to the ones we make when we are feeling calm, loved and supported. The art of persuasion in the hands of a well-meaning person is about making sure that we feel the appropriate response in any given situation. Of course, rhetoric is also very often used by people with malign intentions and, in these cases, it is instructive to note how the misdirection of anger and fear are often used to whip up a mob, trigger a Twitterstorm or to create clickbait.

THE SPEED READ

Rhetoric, the art of seeing how to use a speech or presentation to persuade an audience, is crucial because it is one of the ways of revealing the truth. You can't always persuade someone using only facts, so persuasion is also key. You need a deep understanding not only of logic, but also of character and emotion if you want to persuade people and make effective changes to the world around you as a result.

Enchiridion

Epictetus, second century AD

Epictetus was a Greek philosopher who taught his students that clear thinking and self-reflection are key parts of a good life: philosophy is not just an academic discipline but a practical one.

As a Stoic, he put forward the idea that we should make a calm, unemotional response to events in our lives. But he falls between two extremes in the movement: the early Stoics suggested we should suppress and deny desire and emotion in order to achieve 'eudaimonia' (an even, good temper), while the later Stoics focused on acknowledging emotion and desire without letting them dominate our decisions.

Epictetus sat somewhere between the extremes, arguing that emotion and desire are often simply illogical. He even questions why we are sad if a loved one dies, arguing that it is either selfish (thinking of how it affects ourselves to miss the departed) or a failure to realise that they are in a better place.

His *Discourses* was a summary of his teaching compiled by Arrian, a second-century student of his in the schoolroom in Nicopolis where he seems to have been a kind of tutor for the sons of rich families. He taught three particular fields 'in which people who are going to be good and excellent must first have been trained. The first has to do with desires and aversions – that they may never fail to get what they desire, nor fall into what they avoid – the second with cases of choice and of refusal and, in general, with duty – that they may act in an orderly fashion, upon good reasons, and not carelessly – the third with the avoidance of error and rashness in judgement.'

The *Enchiridion* is a shorter work, summing up some of the key points of his teaching in a book of maxims (the title translates as 'rulebook' or 'manual'). Epictetus urges his pupils to concern themselves with those things in life which they can influence. 'The things in our control are by nature free, unrestrained, unhindered; but those not in our control are weak, slavish, restrained, belonging to others. Remember, then, that if you suppose that things which are slavish by nature are also free, and that what belongs to others is your own, then you will be hindered.'

This is a particularly interesting observation for Epictetus to make, having started his own life as a slave. According to one account one of his legs was once broken deliberately by his master. When Epictetus argues that we can find dignity in all situations and suggests we shouldn't let ourselves be made a slave by allowing someone else to provoke us to anger, he is speaking from a deep well of understanding.

Epictetus also teaches that we are responsible for our own actions, need to practise a high level of self-discipline, and to reflect upon ourselves and our choices. He emphasises the idea that we should always be aware of our own ignorance and gullibility and to accept them with equanimity.

The *Enchiridion* offers a range of advice on how to develop a stoic attitude and to attain apatheia (clear judgment). To live a Socratic life in pursuit of these goals, we need to pursue four cardinal virtues: sophrosyne (temperance and self-control), dicaeosyne (righteousness and truthfulness), sophia (wisdom and common sense), and andreia (courage and a bit of oomph). If we do not indulge desires or wallow in emotion, we can achieve harmony with the world around us and detach ourselves from pointless distractions.

In an age of increasing nationalism, it is interesting to note that Epictetus also had something to say on the subject of attaching your loyalty to such trivial things as your place of origin. 'If what philosophers say of the kinship of God and men be true, what remains for men to do but as Socrates did – never, when asked one's country, to answer, "I am an Athenian or a Corinthian", but "I am a citizen of the world".'

The *Enchiridion* is a useful reminder of the fact that, even two millennia ago, there were teachers like Epictetus who placed clear, intelligent, honest thinking at the heart of their guide to a good life.

THE SPEED READ

There are many things that affect us which we have no power to affect. If you allow yourself to be unduly affected by those things then you are behaving foolishly: instead develop a clear understanding of who you are, don't allow emotion or desire to overwhelm your judgment, and practise self-control, honesty, truth and courage in the pursuit of a good life.

The Story of Philosophy:
A Concise Introduction to the World's Greatest Thinkers and Their Ideas

Bryan Magee, 2001

For the most part the books in this collection are arranged in chronological order, but we're making an exception for this title because it is a look back over the millennia at all the main strands of philosophy.

Bryan Magee was a British philosopher, broadcaster, and author who specialised in explaining philosophical concepts to a popular audience. And while philosophy isn't strictly the same thing as smart thinking, the roots of the latter lie in the former, which is mostly concerned with asking difficult questions about the nature of the world we live in and attempting to resolve them. A basic grounding in philosophy and the different ways it has approached problems of existence is invaluable when it comes to applying critical thought.

Magee takes a chatty approach to the subject but his expertise is undeniable. He begins with the Pre-Socratic philosophers and their attempts to understand the nature of reality. For instance, Parmenides argued that everything in nature is constant and unchanging, while Heraclitus took the view that 'everything is flux. Nothing in our world is permanent. Everything is changing all the time.' (Interestingly, modern atomic science has revealed that both philosopers were essentially correct. The constituent elements that make up atoms are effectively eternal and unchanging, but constantly re-combine, making new elements and substances.)

After working his way through Socrates and the idea of dialectic, Magee moves to the Stoics, Cynics, Epicureans, and Sceptics, before reaching the era of Christianity and noting the ways in which earlier Greek philosophers influenced Christian thinkers such as the Neo-Platonists.

He covers the beginnings of true science and the way that scientists and rationalists such as René Descartes (see p.15), Spinoza and Leibniz undermined Christian philosophy to create the foundations of modern thought. This leads Magee into one of the other great dualities in the story of thought: the battle between rationalists, who believed we could reach an understanding of life, the universe, and everything through pure logic, and the empiricists (including Locke, Hume and Burke) who insisted that we could only do this through scientific investigation of empirical facts.

This also touches on the problem of knowledge, still a difficult subject for philosophers. Without going into too much detail, the problem is that any attempt to define certainty of knowledge seems to be subject to questioning and counter-examples. It feels like a hopeless task at times, especially when you reach the twentieth century, when the attempt to base mathematics on pure logic was undermined by Kurt Gödel's incompleteness theorems, though that's another long story . . . (see p.38)

Magee takes us through the golden age of German philosophy, from Kant to Nietzsche, and in each case he makes a pretty good fist of explaining complicated ideas in simple terms. He looks at the way that the revolutionary French thinkers such as Rousseau and Voltaire helped to create the conditions for a political upheaval, as philosophical theories such as utilitarianism (which bases morality on outcomes rather than intentions or virtue) and pragmatism underpinned the development of modern democracy.

And, finally, he takes us on a whirlwind tour through the twentieth century, looking at thinkers as disparate as Frege and the existentialists. He also has time for Wittgenstein, who analysed language as though it were a mutual game between speakers and came up with memorable images such as the duck/rabbit, which demonstrates how we can perceive the same thing in two different ways, by showing how a specially drawn picture of a duck can be looked at in a different way to appear to be a rabbit.

If you want a shortcut to gaining a broad understanding of how philosophical traditions have developed over the centuries, this is a nicely illustrated, concise and comprehensible guide that covers all the main bases without being glib or simplistic in the process.

THE SPEED READ

A guide to three millennia of philosophical thought, introducing thinkers who have explored the nature of existence, the question of what we can and do really know, and how we have used a variety of philosophical approaches to try to apply smart thinking to the big questions.

The Path:
What Chinese Philosophers Can Teach Us About the Good Life

Michael Puett and Christine Gross-Loh, 2016

Michael Puett is a professor at Harvard, teaching a popular course on classical Chinese ethical and political theory. After the writer Christine Gross-Loh wrote an article about him, they got together to produce this book, essentially a brief guide to Chinese philosophy.

They cover a variety of thinkers who don't always agree with each other, but the book brings out some common strands in which Confucius features heavily. It is an interesting work partly because it offers an alternative approach to the self than that most commonly found in western thought and in self-help. In the western tradition there is a strong emphasis of coming to an understanding of who you truly are and remaining authentic to your 'true self'. By contrast *The Path* focuses our attention on the Confucian idea (also present in much Indian philosophy) that there is no true self. Puett instead describes our minds as 'a messy and potentially ugly bunch of stuff'; we are a bundle of feelings and instincts, without a single guiding force.

Puett has specifically acknowledged how at odds this puts him with the self-help tradition: 'I think of it as sort of anti-self-help. Self-help tends to be about learning to love yourself and embrace yourself for who you are. A lot of these ideas are saying precisely the opposite – no, you overcome the self, you break the self. You should not be happy with who you are.'

If we are a bundle (an idea also found in David Hume, among western philosophers), we are prone to falling into predictable patterns of behaviour. The Confucian response is to disrupt this activity through the practise of ritual: whether it be ancestor worship or paying daily respects to a shrine. (There is a similar idea in Shinto Buddhism: in Japan it is common for people hurrying around the hectic streets of Tokyo to pause at a miniature shrine, lowering their head in thought for a few minutes before clapping and then moving on with their lives.)

While our modern lives may not generally allow for such ritual, the authors suggest alternative ways of disrupting the tendency to get into a rut. For instance, make a small change to daily routines, or a conscious alteration of the ordinary, whether it be asking a shopkeeper how they are, or getting off at a different bus stop. This can help you to step outside of your usual patterns and understand the ways in which they constrain you. Such small human interactions can also boost your emotional intelligence by obliging you to treat another person with respect and to pay attention to their responses and behaviour.

It's been pointed out that the approach recommended in the book also shares a lot of ground with 'virtue ethics', the sum of the innate qualities of a person combined with their desire to live a good life. The idea was commonly explored by writers like Aristotle, but was generally pushed to one side by the

Judeo-Christian tradition, in which adherents were advised that faith was the one requisite quality, and that virtue would result from that faith (practising virtue in isolation from faith was deemed impossible.) However, virtue ethics have been revived by recent philosophers such as Alasdair MacIntyre. The idea is that, through ritual or disruption, we should aim to train our behaviour and emotional responses and this will lead us to becoming more virtuous.

One problem with the book is that in a fairly short space of time several differing writers are covered, and the section on each is fairly lightweight, focusing on a particular issue: from Mencius we learn why plans generally go awry, from Zhuangzi we learn about spontaneity, and from other philosophers we read thoughts about nature, artifice and the nature of power. This can be a bit unsettling, as we are jumping between quite different schools of thought. It also means that we don't get a huge amount of depth when it comes to interesting Chinese ideas such as *wu-wei* (inaction).

The book provides a good example of the sort of hubristic claims that are so often used to hype a book. The authors promise that it will 'flip on its head everything we understand about getting to know ourselves and getting along with other people'. This is definitely overkill, but it is nonetheless an interesting and rewarding read which may inspire you to make some changes in the way you approach life.

---THE SPEED READ--------------------------------

A brief guide to Confucius and other early Chinese philosophers: we are a bundle of qualities and emotions – there is no true self. True meditation and mindfulness are built around coming to this understanding, but we can also use ritual and disruption in our daily lives to reveal and understand the patterns we unwittingly fall into and to develop more virtuous ways of doing things.

Discourse on Method

René Descartes, 1637

The origins of 'smart thinking' lie in philosophy. While we've avoided including too many pure philosophy books in this collection, a few titles seemed worthwhile for the role they have played in provoking readers to deeper thinking.

Discourse on Method is a fundamental title for western philosophy, a fascinating example of rationalism (essentially, the attempt to build up a system of knowledge using logic alone) and the ultimate source material for the *Matrix* movies, to boot. The author can be quite funny, at least in the early stages of the book, writing, 'Good sense is, of all things among men, the most equally distributed; for everyone thinks himself so abundantly provided with it, that those even who are the most difficult to satisfy in everything else, do not usually desire a larger measure of this quality than they already possess.'

Descartes, a brilliant mathematician and a religious believer, was bothered by the problem of scepticism, the fact that almost everything we think we know seems like it is susceptible to reasonable doubt. He decided to design a method that would help him to explore this and, hopefully, to attain certain knowledge about the world. One of his four principles is the most important for our purposes: 'The first was never to accept anything for true which I did not clearly know to be such; that is to say, carefully to avoid precipitancy and prejudice, and to comprise nothing more in my judgment than what was presented to my mind so clearly and distinctly as to exclude all ground of doubt.'

Given this, he vowed to suspend his belief in every single thing that he could imagine being untrue: he gives a model demonstration of the method of doubt, pointing out all too persuasively that almost every one of his beliefs could be wrong. He takes this to such an extent that he decides not to trust his perceptions, reasoning that he could be the prisoner of a demon who has placed him in a vat and is somehow feeding him those apparent perceptions. (I did flag up the comparison with *The Matrix*, didn't I . . . ?)

At this stage, all seems lost. But then he sees a tiny ray of light in the darkness. Even if all his perceptions are false, they can only exist because he is thinking. This leads him to one of the most famous lines in all philosophy: 'I think, therefore I am.' (I could get bogged down in the many philosophers who have pointed out flaws in this approach, as it assumes that such a thing as the self exists, even in this simple step, but let's not get distracted by that just now.)

From here, things start to get a bit shaky. Descartes suggests that the method of doubt can't pass judgment on reason because it is actually based in reason. And while your head is still spinning trying to work that one out, he slips in the idea that there must be a God, as he is the only guarantee that reason isn't being tricked. He offers a few proofs for the existence of God, including the ontological argument (which is, essentially, that God is the greatest being conceivable and, since we can conceive of him, he must exist). Then from there he goes back to assuming that his perceptions are trustworthy, since God wouldn't allow him to be tricked, and he starts the complex process of trying to build up a system of knowledge about maths, physics and the world using his perceptions as foundations.

This section of the book is interesting, but the part that is best remembered – and which made the work a classic – is its initial journey into darkness and doubt, and the first few steps he attempted to take in order to resolve his moment of absolute uncertainty. In terms of smart thinking today, it might be worth bearing in mind that a scaled-down version of the method of doubt can be a useful thing to deploy in many situations. You needn't go so far as imagining a demon is tricking you to take a few moments to think about all the things you might have wrong in a situation and what the consequences could be.

THE SPEED READ

I want to build a complete, reliable foundation of knowledge, but everything seems open to doubt. Even my perceptions might be the result of an evil demon's trick. But wait a moment: I know I am thinking, so I must exist. And if I exist, God must exist and, er, he wouldn't lie to me so . . . hurrah, my perceptions are trustworthy after all!

The ABC of Relativity

Bertrand Russell, 1925

I was exploring my parent's bookcases one day when I was about twelve, and I stumbled upon a book that looked like a science-fiction novel.

It showed the swirling surface of the sun along with some mathematical symbols and equations. On further exploration it turned out to be a beginners' guide to Einstein's theories: Bertrand Russell's *The ABC of Relativity*. While I certainly didn't understand large parts of it, it made a deep impression on me, to read such astoundingly deep theories explained in such a chatty, witty manner.

In the years after World War I, Russell had found himself short of money as, in spite of his academic reputation as a brilliant philosopher (especially in the field of mathematical logic), he didn't have an academic appointment. However, that wasn't the only reason he turned to writing books: he also had a deep conviction that the horrors of the war had been partly caused by ignorance and prejudice and believed that teaching people about the wonders of the universe and philosophy was a way of making the world a better place.

This drive is one of the things that gives this book such focus and humour, although the humour is sometimes quite dark – for instance when Russell muses about atomic weapons. Having mentioned that there is a limited supply of uranium from which such weapons (that hadn't then been perfected) could be built, he reassures the reader that there is no shortage of hydrogen for hydrogen bombs, so 'there is considerable reason to hope that the [human] race may put an end to itself, to the great advantage of such less ferocious creatures'.

The book has aged remarkably well as a basic guide to relativity, although it predates Hubble's discovery of the expansion of the universe and raises some questions that have since been resolved. The physicist Felix Pirani updated later editions, and the most obvious anachronisms have been removed, although there is a certain charm to such old-fashioned talking points in the original version as 'aether'.

Russell starts the book with a look at the ways in which we perceive the earth and heavens. He asks the reader to imagine an observer who is drugged and put into a hot-air balloon: on awakening, the drugs have wiped his memory. When he looks down at the Earth, he is reliant on his sensory perceptions for his idea of what he is seeing. It is night-time on the fifth of November over the UK, fireworks night, and all he perceives is a darkness occasionally lit up by showers of bright lights. He would presumably conclude he is looking into a void in which bright objects occasionally come into existence and then disappear. The point of this is to focus our attention on how we perceive the heavens and how little information we actually have. It is also to show the unreliability of our everyday assumptions about what we are perceiving and what the objects around us 'are'. We need to put them to one side, especially if we are to start thinking about Einstein's amazing theories.

His account of those theories is remarkably concise and clear. He talks about the fact that there is no single point of reference from which motion is measured, meaning that all things are in motion, relative to each other. He gives a clear explanation of why this means conventional physics and the geometry of Euclid fail to apply. He talks about the difficulty of deciding whether distant events happened 'before' or 'after' each other, and shows that time and mass are only consistent if you assume they are being measured by an observer at the same velocity.

He gives a simple explanation of the relationship between light and gravity, and the way that light bends, and goes on to explain that Newton's concept of 'forces' is actually merely a way of describing something different: gravity is actually more like a hill, and matter or light will tend to take the easiest path from A to B. All of this makes perfect sense to the reader, although it can be hard to come away from the book with a lasting understanding of the theory. (However, this is a problem for most readers with any account of relativity, no matter how well explained it is: it is simply too mind-boggling and too remote from our everyday perceptions of reality to truly sink in.)

Russell is also interested in the philosophical implications of relativity. At the time, there was something of an over-reaction to the concept, which was often felt to prove that 'everything is subjective': Russell calmly points out that this was something of a confusion. When Einstein talks of an 'observer' and the different perceptions that the observer will have depending on velocity and position, we don't need to imagine the observer is a conscious entity: it could equally be a photographic plate recording the arrival of a beam of light. Relativity is merely a way of understanding the physical universe, and the philosophical importance of it can be overstated.

Russell concludes the book with an interesting meditation on power, musing about the abstract power a financier shows when he buys and sells goods that he hasn't seen or touched, and comparing this to the power that can be bestowed on someone by abstract mathematical knowledge. Similarly, the theory of relativity and a knowledge of atomic theory are mighty tools. 'The final conclusion is that we know very little, and yet it is astonishing that we know so much. And even more astonishing that so little knowledge can give us so much power.'

THE SPEED READ

A remarkable guide to Einstein's theories by another of the most brilliant minds of his generation. The mathematical philosopher Bertrand Russell, writing for a popular audience, calmly and concisely takes the reader through the basics of relativity, and this is still a better primer than many of the books that have appeared since with titles along the lines of *The Fifteen-Minute Guide to . . .* or *Einstein Made Easy*.

How to Read a Book:
The Classic Guide to Intelligent Reading

Mortimer J. Adler and Charles Van Doren, 1940
(revised 1972)

You probably think you know how to read a book, right? Just open it and start reading! Well, there is a lot more that can be said about it and this is a classic guide to the process.

Mortimer J. Adler was a philosopher and educator who was on the editorial board of Encyclopedia Britannica, in particular their 'Great Books of the Western World' programme. This makes him a heavyweight guide.

His book was first written in 1940, with a 1972 edition revised with the editor Charles Van Doren. Adler takes a very high-minded view of the purpose of reading, writing, 'A good book can teach you about the world and about yourself. You learn more than how to read better; you also learn more about life. You become wiser ... in the sense that you are more deeply aware of the great and enduring truths of human life.' He dismisses the possibility that other sources of 'amusement and information' can serve a similar purpose, describing them as artificial props. He also devotes some time to criticising common educational methods, using anecdotes from his life to point out their shortcomings.

The essence of his approach is that reading a book involves a deep, serious conversation with its author. He acknowledges that books can provide entertainment and information, but for him the most important role they provide is to endow the reader with understanding. This is partly because reading a book, especially a 'difficult' book, challenges the reader to make cognitive leaps, to acknowledge complex or discomforting truths and to see the world in a different way. He writes, 'We must know how to make books teach us well.'

He makes the case that understanding is best communicated by the person who first acquired that understanding. This is his argument for placing a strong focus on the 'great' books (the appendix includes a stentorian reading list that will make most readers feel very small when they contemplate how few they have read), although the arguments and methods in the book can be used for literature in general.

In order to be in sufficient communion with an author, you first need to be able to understand the basics: what is the book about? What are the detailed elements that make up the whole? Is it true, or partially true? And what is your response to it? Throughout, the author is stressing the idea of reading as an active activity, in which your responses are as important as merely absorbing what the author is saying.

Adler moves on to higher levels of reading. Firstly, he gives a detailed picture of what he calls 'analytical reading'. This can be broken down into three aspects (although this needn't imply reading the same book three times . . .): structural, interpretative, and critical.

Structural reading involves the reader getting an understanding of the overall makeup of the book: how it is divided up, what the logic of the ordering is, what problem the author is trying to solve, how they set that problem up and so on.

Interpretive reading involves identifying the logic of the author's arguments, identifying any particular concepts or phrases that the author uses, and noting where the various stages of this argument are made. Throughout this process Adler suggests annotating your edition as you read: 'Full ownership of a book only comes when you have made it a part of yourself, and the best way to make yourself a part of it – which comes to the same thing – is by writing in it.'

The third level of reading involves critiquing the book: 'You must not argue with a book until you fully understand what it is saying.' But, having understood the structure and how the author is setting up a problem and solution, the reader has put themselves on a level with the author, and can now feel entitled to engage in questioning whether or not they accept the author's thesis.

Up to this point, this might sound like quite a formal description of what naturally happens when you read a serious book. However, Adler also considers a broader view of reading. He mentions, for instance, the additional depth that can be brought to the great books if you also read the authors and books that influenced them in turn. And he discusses syntopical reading. This involves a reader coming to a deeper understanding by reading different books on the same subject. And if that seems obvious, the detailed approach he suggests is quite exhaustive.

First, he suggests researching your bibliography for a particular subject, identifying the classic texts that have dealt with the same theme, and narrowing down your comparative texts to a shortlist, having briefly skimmed the content of a longer list. Then you need to revisit the theme you are exploring, in case it has to be reformulated and, finally, you can produce a final list for intensive reading.

At this stage, you are not reading each of these books in depth, just focusing on the theme you are exploring. And you are attempting to find the most relevant passages, to understand them in terms of the way you are formulating your theme, and to be clear about what each of these authors is saying.

The idea is to broaden out your investigation of an idea to encompass the entire field of literature, to identify the crucial players in that field and to create a mental map of how the great writers have dealt with a particular theme. In the end, this may seem highly academic and beyond the scope of the average reader, but it nonetheless provokes you into thinking about what is happening when you read any book: at some level, you have a mental map of how different people think about a sub-ject. Even if you can't go to Adler's extremes in researching and reading syntopically, you can take a simpler message: that you will have a deeper understanding of an author if you not only critique their ideas but place them and interpret them in a much broader context. And therein lies the value of the book, which will at the very least make you want to become a 'better reader'.

┌─**THE SPEED READ**─────────────────────────────────┐

A serious but accessible guide to reading books (and especially Great Books) with the aim of attaining greater understanding and wisdom, and how this can help you grow through life: 'If the book belongs to the highest class – the very small number of inexhaustible books – you discover on returning that the book seems to have grown with you.'

└──┘

Man's Search for Meaning

Viktor E. Frankl, 1946

Man's Search for Meaning describes the experiences of Viktor Frankl, an Austrian psychotherapist, in Nazi concentration camps including Theresienstadt, Auschwitz, Kaufering III and Türkheim.

One of his main conclusions was that a positive attitude had a huge impact on whether or not a prisoner survived the camps: in particular, he argued that those prisoners who maintained a positive visualisation of their future would have a better outcome. He applied this insight in the psychotherapeutic method he developed, logotherapy. This involved encouraging the patient to find a purpose in life about which they could feel positive, and then to imagine that outcome in an immersive way.

This makes *Man's Search for Meaning* a fascinating book in the development of the whole idea of 'positive thinking', which itself has had a huge impact on self-help literature, especially in the USA. (The book has been named as one of the country's ten most influential books and has sold over twelve million copies.) Some psychotherapists see this impact as malign, that it gives people false, 'magical' ideas about the possibilities offered by their future, and encourages them to feel disheartened on those occasions when thinking positively just isn't enough.

Referring to incidents from his own experiences in the camps, Frankl discusses the psychological reactions of prisoners to their circumstances. Many reacted with shock in the initial stages, turning to apathy as they became gradually institutionalised, and finally feeling depersonalised and bitter, sensations that could persist into their lives after they were finally freed. Frankl devotes a significant part of the book to his observations of the difficulties prisoners faced in adjusting to life after the camps.

One of Frankl's conclusions was that each minute of our experiences shows us the 'meaning of life'. He gives the example of a hunger strike the inmates participated in to protect a fellow prisoner: even in such an awful situation, many inmates could feel that they were trying not to let someone down, whether it be God, or someone close to them. This leads him to observe that we always have the freedom to choose to act on such feelings, and it is partly hope for the future that encourages us to hold on to our spiritual self and sense of morality.

He also argues that there are only really two races: people who are decent and people who aren't. He points out that there were decent Nazi guards who didn't hate or torment the prisoners and there were also prisoners (for instance the *kapos* who assisted the guards) who didn't behave in a decent way.

There are two lines of objection to Frankl's arguments. Historians have shown that his account of his time in the camps is not reliable: for instance he was only at the death camp Auschwitz very briefly, before being transferred to the work camp Kaufering III, but he blurs this issue, suggesting that experiences from the latter actually happened in the former. And he has been criticised for his willingness to perform lobotomies on Jewish prisoners before his own arrest, possibly in an attempt to ingratiate himself with the Nazi regime. Frankl has also been criticised (for instance by the writer Lawrence L. Langer) for the boastful, self-centred tone of the book and a somewhat inhumane element in the way that Frankl treated his fellow prisoners as subjects to be studied.

But the most significant criticism may be of the idea itself that a positive attitude helped prisoners to survive the camps. Historians haven't found any way to support this, so at best it is a conjecture. And it is a conjecture that is in danger of implying that the victims of the Holocaust should be blamed for just not being positive enough.

This is a general problem with the cult of positive thinking. Happy-clappy books such as *The Secret* (Rhonda Byrne) and *The Power of Positive Thinking* (Norman Vincent Peale) have been hugely successful in communicating the message that you can become rich, healthy, successful or whatever, simply by being positive enough. The latter book was a big influence on those, such as President Trump, who seem to believe that any negative, inconvenient facts about the world can be simply wished away. And all such books are keen to convey the idea that any poverty, illness or accident can be overcome, and if not, it's your fault for not being positive enough. Positive-thinkers can file you away with those Auschwitz victims who weren't sufficiently upbeat about their fates, and ignore you rather than having sympathy for you as someone who just got dealt a bad hand in life.

Frankl's book is in many respects interesting, especially in its accounts of the psychological experiences of liberated prisoners, but it is perhaps best regarded as a flawed, unreliable account that has a potentially unpleasant side to it, and left in the past where it belongs.

THE SPEED READ

A psychotherapist who spent time (though not as much time as he implies) in the death camp at Auschwitz details his observations of the mental state of his fellow prisoners both during the time in the camps and after liberation. His ultimate suggestion, that a positive attitude was necessary for survival, is historically dubious and potentially quite unpleasant in the way it implies the victims who died were partially responsible for their own deaths.

The Use of Lateral Thinking

Edward de Bono, 1967

Edward de Bono's first book, *The Use of Lateral Thinking*, was a huge bestseller in the 1960s and 1970s at a time when books on creative thinking were not as common as they are today: to a degree this book invented a new genre all on its own.

This work introduces de Bono's concept of 'lateral thinking', essentially a way of escaping the rigidity of mainstream, logical thinking. He describes it in glowing terms as a habit of mind that can be acquired by anyone with a bit of practise. He sets it up in contrast to the status quo of 'vertical thinking'. This is logical thinking that sticks to the most likely path: he explains the vertical thinker will dig a hole in the one most promising place, whereas the lateral thinker might dig several holes in different places and see what happens. This idea of moving on to a slightly unexpected solution – a kind of 'a-ha' moment – is the core of the book: he suggests that after a solution has been found through lateral thinking, we will often see the original problem in a completely different way.

When it comes to creative endeavours, de Bono argues that vertical thinking is often useless and actively prevents the development of new ideas. Going back to digging that hole, if we put all our efforts into completing the digging of a hole we have already started we will never find the more interesting thing that is buried just a few yards away. Rather than sticking to one way of thinking about a problem, we need to try as many different ways as possible to propel us to new solutions and possibly a new view of the problem itself. Sometimes this might involve leaving some things to chance, to find out what happens if we don't interfere in a situation.

There are some fascinating examples in the book, which often convey de Bono's point more effectively than his regular repetitions of how wonderful lateral thinking is. He talks about the way that, when we describe and name something, we are imposing a label and subjective view on it. He includes an exercise in which shapes are described by viewers to demonstrate how different people might see the exact same set of lines.

He also talks about brainstorming, and about magpie minds that wander and wonder, and store bits of ideas and images up for later. Then, when brainstorming sessions take place, the lateral thinker with a magpie mind has a whole stock of things that might have seemed irrelevant or pointless in isolation, that they can bring to bear on the problem under consideration. This is illustrated with a neat experiment: if you want to make a chain of paper clips, you can carefully attach a series of them together, one to the next. Alternatively, you can open up each paper clip slightly then toss them around in a pan, and a chain or several lengths of chain will naturally result as the clips come together. The latter process is more like evolution or the random idea generation that is key to lateral thinking. It will also produce a design which is more 'interesting' and complex than the chain that results from a vertical thinker threading clips together.

Of course, in order to use lateral thinking, one sometimes needs to start out by identifying the dominant ideas that vertical thinking would produce and to recognise their boundaries. This helps to loosen things up and move things in a different direction. De Bono gives several examples of ways you can change the obvious 'rules' of a situation to apply lateral-thinking techniques.

In a sense the basic theme of the book is very simple: if you look at any problem (and its most obvious solutions) as a pattern, then you need to break up or reorganise that pattern in order to find a less obvious but potentially better solution. No idea should be immediately discarded, no matter how extraordinary or silly it may seem. You can allow yourself to make mistakes during the process, as these may be the grit in the oyster's shell that produces a pearl of an idea at the end of the day.

The two criticisms most often made of the book are that de Bono's self-congratulatory tone can grate, and (more seriously) that there is little evidence to support some of his claims. But as a source of inspiration when it comes to trying to find creative solutions it is still effective. It has been followed by many other books analysing the creative process, but it holds an interesting place in the history of books about smart thinking.

THE SPEED READ

Vertical thinkers look for the most obvious place to look for a solution, then keep digging no matter how long it takes. Lateral thinkers dig holes in random, silly or unexpected places and can find unexpected but brilliant solutions as a result.

Lessons of History

Will Durant and Ariel Durant, 1968

Will and Ariel Durant were American polymaths who wrote the monumental eleven-volume series *The Story of Civilisation*. Intended as an accompaniment to that series, *Lessons of History* boils down some of the enduring lessons that can be learned from humankind's past into a single volume.

It is an impressive attempt to condense periods and explain the cyclic trends that have repeated throughout history, without taking a judgmental view of any political system or type of civilisation. Does it succeed? Well, it is impossible to write a book like this without some biases showing, but it is a truly interesting read.

The authors make fascinating comparisons over broad stretches of time: they compare the Greeks of Plato's time with the French of recent centuries, and the ancient Romans to the English, on the basis that 'motives and ends remain the same: to act or rest, to acquire or give, to fight or retreat, to seek association or privacy, to mate or reject, to offer or resent parental care'.

To give a brief idea of the scope of the book is difficult, but here are a few summaries of its main points. The authors focus on how biology and history have interacted: the largest part of history involved groups or tribes in competition with each other, fighting and killing in order to survive and prosper, and biological selection has been a constant feature in the way that societies develop. The authors describe freedom and equality as opposites on a spectrum, given that perfect freedom would entail extreme inequality, while perfect equality would be a constraint on freedom. Their thesis is that the individuals who are below average in terms of biological selection are more likely to be the ones who want equality, while it is generally the stronger individuals who desire freedom and who succeed. They explore past societies in which some form of socialism has been attempted – for instance, in China under Wang An-shih (1068–85 AD) the state owned most industries including agriculture, and provided a kind of safety net or welfare for the working class. The system eventually collapsed due to high taxation and corruption in the bureaucracy. The Durants' conclusion is that utopias of equality are not sustainable.

When it comes to religion, they identify a tendency of puritanical movements to grow in a society when the laws are weak and morality is left up to the individual, whereas scepticism and pagan thinking rise when the state is more tyrannical (and the church is less powerful as a result). They speculate that many sins were once a virtue: since pugnacity, greed, and brutality were advantages in the original struggle for survival.

There are interesting chapters on the role of money and the economy in history: great monuments are built by states and dynasties with full coffers. The crusades were driven by a western desire to control valuable eastern trade routes. The Renaissance was rooted in the economic power of the Medicis, and the French Revolution fundamentally came about because the middle classes had gained economic power. But in all these situations it was true that 'the men who can manage men, manage the men who can manage only things, and the men who can manage money, manage all'.

They treat war as an inevitable feature of civilisations, pointing out that there have only been a couple of hundred years in the last few thousand without wars. They predict that war will only be overcome if one great power wins such a decisive victory over its rivals that it can dominate the world as the Romans did in Europe and northern Africa at their height.

Finally, the authors take a pretty cynical view of progress and the inevitable rise and fall of civilisations. They comment that the decline of a civilisation is generally rooted in the failure of its political and intellectual class to deal with its challenges – this sounds worryingly prescient in the present day. They suggest that there has been little actual progress throughout human history: we may have new technologies, but they are only used to pursue the same old goals, such as the acquisition of material goods, the pursuit of sexual gratification, competition and military conflict. And while they try hard not to judge different eras, they do allow themselves to ponder whether or not the Middle Ages and Renaissance might have been wiser periods than the current day: those were times when there was less focus on science and power and more on art and mythology. In the authors' view, the result may have been societies in which more individuals led balanced, fulfilled lives.

THE SPEED READ

The beauty of this book is that it brings out some of the echoes and parallels across historic eras and provokes the reader to consider what the underlying driving forces of civilisation are (as opposed to taking much shorter views of political and economic history). Written in a time before 'big history' became a trend, this is truly a panoramic overview of all of civilisation, and you needn't agree with everything the authors believe to admire their breadth of vision and ambition.

Gödel, Escher, Bach

Douglas R. Hofstadter, 1979

Gödel, Escher, Bach is a dazzling masterpiece that blends genres and styles to link ideas about mathematics, music, art, computer science, and the nature of consciousness itself. It is told through a mash-up of puzzles, dialogues, essays, and self-referential pranks that makes it unique. It has also helped many people to get a handle on some truly complex ideas.

Hofstadter begins by discussing the composer Johann Sebastian Bach's canons and fugues (essentially recursive figures in which a single theme is repeated against itself, with fugues the looser and more complex of the two forms). Bach's *The Musical Offering* includes a six-voice fugue (bear this is mind, as it is important later on). The piece is an unusually complex compos-ition that includes an endlessly rising canon, a figure that can be played in a loop, as it cycles back to the beginning. Hofstadter refers to this as his first instance of a 'strange loop'.

The book is structured so that themes are continually revisited in a playful way during dialogues in which characters such as Lewis Carroll's Tortoise and Achilles have playful conversations relating to the text. We move on to other examples of strange loops, starting with the Epimenides paradox, in which it is impossible to say whether the statement, 'This statement is false', is true or not. It was pondering such paradoxes that led the philosopher Bertrand Russell (see p.18) to formulate his own paradox. His 'set of all sets which are not members of themselves' is an impossibility and the existence of which may seem an arcane, theoretical issue. But the mathematical logic at the time attempted to define everything in terms of sets. So, Russell's paradox actually led to the conclusion that it is impossible to establish a consistent set of founding principles (or 'axioms') on which mathematics is based.

The troubled philosopher and mathematician Kurt Gödel took Russell's work further when he formulated his 'incompleteness theorems', a more fully formed demonstration of the problem. In any attempt to build a mathematical system from foundational axioms, there will be statements that are true but unprovable, and no axiomatic system can demonstrate its own consistency.

The common culprit in all these mathematical conundrums (and 'strange loops') is self-reference. Similar problems arise in the impossible etchings and illustrations of M. C. Escher, whose artwork is heavily featured in the book. His famous endlessly ascending staircase is another example of a strange loop, as is his image of two hands drawing each other.

Hofstadter also introduces some delightfully silly additional examples such as 'Record Player X', which destroys itself if it plays a vinyl record titled *I Cannot Be Played on Record Player X*. He connects loops like these to Zen koans, paradoxical sayings used to provoke a state of confusion and doubt in the student. The idea is that such loops and koans allow the student or reader to step out of reality by revealing the inherent inconsistency and confusion.

A similar role is played in the book by puns and by Escher's illustrations that show, for instance, an interweaving pattern of what might be birds or frogs depending on the interpretation our brains make. Seeing two options involves us finding ways to 'jump out' of our current mindset to see the world differently.

Hofstadter moves on to a fascinating examination of the role of self-reference and recursion in computer programming languages. Some readers may start feeling a little weary by this stage of the book, as the endless wordplay, puzzles, and puns can be distracting when the mind is boggling at such a complex array of ideas and the interactions between them. And at times Hofstadter's determination to dazzle the reader makes one wonder if there is a bit of a Wizard of Oz thing going on, with the smoke and thunder distracting us from the real trick being played out behind the scenes.

It is worth persisting with the book, as Hofstadter's ultimate goal is to examine the way in which high-level systems can be explained at the fundamental level by lower-level rules, just as with complex computer programs. The question is, inevitably, whether or not the human mind can truly be compared to a machine or program. Referencing the Turing Test – in which artificial intelligence is measured by whether or not a human can tell they are talking to a machine rather than another person – and many other examples of real or possible programs, Hofstadter ends with some fascinating meditations on the idea that strange loops and self-reference are keys to understanding what consciousness truly 'is'.

You needn't accept all of his conclusions to have had a giddy, enjoyable ride with the book. And the final flourish of a six-voice dialogue (see, I told you . . .) involving Achilles, the tortoise and crab, Hofstadter, Charles Babbage and Alan Turing leaves us in no doubt that the author is, at the very least, on top of his game.

---THE SPEED READ---

'Strange loops' are recurring themes found in the work of the composer Bach, the artist Escher and the mathematical logician Gödel. The latter's incompleteness theorems demonstrated the limitations of any attempt to produce a formal basis for mathematics. The underlying problem (which is self-reference) is also significant in computer programming and in the problem of human consciousness. The discussion is accompanied by poems, jokes, musical puns, philosophical flourishes and flights of fancy. Genius.

A Whack on the Side of the Head

Roger von Oech, 1983

This is a venerable book on creativity that has stood the test of time well. Von Oech got a PhD from Stanford in the history of ideas, a discipline he invented himself.

He went on to coach creativity in a variety of ways, as well as enjoying a successful career as an inventor of toys. The Creative Whack Pack, marketed as a portable version of his workshops, is a parallel set of cards intended to stimulate creativity, has sold over a million copies and often comes with the book.

Von Oech's book begins with a brief questionnaire entitled 'Mental Sex' – presumably to make it sound more exciting. There is a tendency in the book to use gimmicky titles which are, like the original illustrations, somewhat irritating. The aim of the test is to show readers that they are more creative than they think. The author talks about the sorts of problems he might set in a creativity workshop – for instance, imagine a corporation has over-ordered ball bearings and is looking for a way to use them: what would you suggest?

Van Oech describes the creative mind as one that wants to 'know it all', acquiring knowledge like a magpie, not knowing when various ideas might come together to create something new. The example given here is that of Johannes Gutenberg, who brought the wine press (that exerts downward force) together with the coin punch (that imprints a design in metal) to create the printing press. It was his knowledge of and understanding of the previous inventions that allowed him to make this creative leap.

Van Oech suggests that 'the hallmark of creative people is their mental flexibility. They are able to shift in and out of different types of thinking depending on the needs of the situation at hand.' This might involve being open and receptive, being playful, or being dogged and self-critical. He characterises four types of thinking via the roles of explorer, artist, judge, and warrior. In The Whack Pack, you choose a card from each of these roles to give your mind new directions.

Von Oech discusses ten 'mental locks', ways in which our ways of thinking obstruct our own creativity, in ten chapters and there's a degree to which the chapter titles are self-explanatory: 'The Right Answer', 'That's Not Logical', 'Follow the Rules', 'Be Practical', 'Play is Frivolous', 'That's Not My Area', 'Don't Be Foolish', 'Avoid Ambiguity', 'To Err Is Wrong', and 'I'm Not Creative'.

His discussion of these locks is often enlightening. For instance in 'The Right Answer' he discusses the way that we often learn that there is one correct way of doing things and that other approaches are somehow improper: obviously this can stifle creativity, so we need to find ways to shake up our assumptions and see new possibilities – such as 'whacks on the side of the head'.

When it comes to the 'That's Not Logical' lock, the author talks about two types of thinking: soft thinking and hard thinking. The former is like a floodlight, which lights up a whole area, so you can see the similarities and connections between different things. The latter is more like a spotlight that allows you to examine something more closely to see its unique qualities. When we are being creative there is an imaginative phase during which soft thinking allows us to link different ideas, and a practical phase in which we need hard thinking to figure out the specifics of a problem.

A large part of von Oech's message is summed up when he writes, 'Every child is an artist. The problem is how to remain an artist after growing up.' As he discusses the remaining mental locks he emphasises the importance of allowing yourself to be playful, to do things that are outside of your comfort zone, to allow yourself to do things that are foolish and perhaps even bewildering. He repeatedly returns to the idea that it is fine to make mistakes and that we can all be creative. Along the way he provides numerous exercises designed to drive his message home and draw out the reader's creativity.

It might be cheesy at times, but it's hard not to be beguiled by the sheer enthusiasm and ebullience of the man: there aren't many books on creativity that are still in print in their fourth decade, and there are good reasons why this is one of the few that is.

---THE SPEED READ---

Sometimes, when you are challenged to come up with an idea, you might think it isn't something you can do, or that you just aren't creative. What you need is a whack on the side of the head, something that shakes up your assumption and releases you from your mental locks, such as the fear of making mistakes, the idea that there is a single right answer to any problem, or the aversion to looking like you are just playing around and being foolish. Once we've unlocked the locks, we can all be creative in the same way as a child.

Chaos:
Making a New Science

James Gleick, 1987

There aren't a huge number of books on complex mathematical subjects that truly capture their subject while appealing to the non-mathematical reader. *Chaos* was a bestseller largely because Gleick succeeded in describing some awe-inspiring mathematical concepts in a readable, engaging way.

There are certain dynamic systems in the real world, including the motion of water in a river or the ways in which weather systems develop, that have long been seen to be somewhat 'chaotic' or unpredictable. What wasn't truly realised until the last few decades is how this connects to the different mathematical area of fractals.

In his 1967 paper 'How Long Is the Coastline of Britain?', Benoit Mandelbrot investigated the quality of 'roughness' that shows, when you measure a coastline, you get different results depending on how close you zoom in. Measure the perimeter of a smooth, coastal road and you end up with a shorter distance than if you were to measure around every cove and promontory. The distance would be longer again if you measured around every rock and pebble of the shoreline.

Mandelbrot realised that this phenomenon could be observed in many natural environments: plant structures, ocean waves, blood vessels, and frost patterns. There was often an element of self-similarity in these structures – the shapes you observe at one level of magnification are repeated as you zoom further in. Such patterns are called 'fractals' – they have many extraordinary qualities, and can be quite beautiful as demonstrated in the beautiful Mandelbrot set, one of the most attractive mathematical constructions conceivable.

Gleick succeeds in drawing the reader into the magic of such ideas partly by focusing on this very visual way of thinking about fractals. He introduces other visually fascinating constructs such as Julia and Fatou sets, known as 'lace' and 'dust' respectively. He connects these with another, well-known mathematical wonder, the 'butterfly effect'. This in turn is based in Lorenz attractors, which are technically a set of chaotic solutions to a mathematical function called the Lorenz equation. The attractors not only look like a butterfly but they also describe a system in which it is impossible to predict the outcome unless we have a perfect knowledge of the starting conditions. For instance, unless we know everything that could affect a weather system, down to the distant flapping of a butterfly's wings, we cannot perfectly predict the outcome of that system. The same applies to many other situations, including traffic systems.

These ideas collectively make up 'chaos theory', a way of describing the mathematics of dynamic systems that are highly sensitive to initial conditions. They were traditionally deemed to be too hard to capture or even describe using maths, and Gleick successfully celebrates the wonder felt by those mathematicians who started to realise that the seemingly arcane subject of fractals was actually the key that unlocked our understanding of such systems.

One reason why chaos theory only really advanced from the 1960s onward is that it is heavily dependent on the computers that became more widely used at that time. They were able to generate the repeated iterations of equations that make up the beautiful fractal patterns and sets. The key thing is that these systems behave as though they are non-deterministic, when in fact they are deterministic. Mathematician and meteorologist Edward Lorenz once summed up the theory this way: 'Chaos: when the present determines the future, but the approximate present does not approximately determine the future.' And being able to use computers to model those systems has allowed us to get increasingly detailed insights into what is actually going on in dynamic systems.

Chaos theory has gone on to find relevance in cryptography, biology, robotics, and much more besides. The way in which Gleick acclaimed it as a major breakthrough in 1987 has been justified over the decades: it is now well-established as a key part of the mathematical canon. But the wonder of this book is the way that Gleick draws you into such difficult ideas using the aesthetic and even linguistic fascination of ideas such as 'strange attractors'. As a result, you barely even notice when he shifts your attention onto the real maths behind the beauty.

THE SPEED READ

A brief introduction to chaos theory, a branch of mathematics that can be used to describe previously unpredictable phenomena such as weather systems, traffic grids and the motion of a droplet of water. A clear, beautiful account that will make you wonder at the complexity that can be conjured up from reiterating a few simple equations.

A Brief History of Time

Stephen Hawking, 1988

Famous for being the 'most popular book never read', *A Brief History of Time* is a classic prop for anyone who wants to look serious and intellectual, whether it be in a café, or through a copy artlessly displayed on the coffee table.

However, in spite of its reputation, it is a rather enjoyable read that can make the reader simultaneously feel smarter (when one of Hawking's simple explanations makes perfect sense) and stupider (when the reader's head starts to spin with quarks, anti-quarks and the likelihood of an infinite cosmos).

Hawking begins with an anecdote previously recounted by Bertrand Russell. An old lady told him, 'The world is really a flat place supported on the back of a giant tortoise.' When he asked what the tortoise was standing on, to show her the folly of her thinking, her reply was, 'You're very clever, young man, very clever. But it's turtles all the way down.' It is characteristic of Hawking's deft touch that this leads into a whirlwind guide to the history of cosmological thoughts, from the ancient Greeks' understanding that the world was probably round through to the most recent theories of the universe.

In relatively short order, Hawking explains his thoughts on space and time, the nature of the universe, particle physics and how the universe started. Throughout he manages to convey both childish wonder ('I am just a child who has never grown up. I still keep asking these "how" and "why" questions. Occasionally, I find an answer'), and the sheer ambition of his thinking ('Our goal is nothing less than a complete description of the universe we live in').

It is interesting to note that Hawking nearly cut one of the book's most famous lines, when he wrote that, should humankind discover a unified theory of the universe, 'We would know the mind of God.' Hawking hinted at his atheism when he said, 'An expanding universe does not preclude a creator, but it does place limits on when he might have carried out his job.' The fact that he was able to even entertain such mystical thought and allow a conceivable place for a creator in his universe clearly expanded its popular appeal, which perhaps accounts for why the quote is so well-known. In truth, Hawking's account of where the universe might have come from is pretty heavily based in physical forces and objects rather than in metaphysics or spiritual ideas.

One reason why readers struggle to finish the book is that later chapters are devoted to some truly mind-boggling trains of thought about the most speculative ideas of modern physics. We may be familiar with the general concept of 'wormholes' from *Star Trek*, but the theoretical detail is a lot harder to grasp. The same goes for superstring theory, which involves trying to think about the world in far more dimensions than our minds generally operate in.

Of course, one can now also read much erudite criticism of the book's areas of oversimplification (for instance, the speed with which he glosses over the idea of entropy and uses it to explain why we experience time in a 'forwards' direction). Such nitpicking is not particularly important for a book like this, one that really isn't aimed at a specialist audience. Hawking always intended to write something that would sell in airport bookstores and – rather to his and his publishers' surprise – he succeeded in creating one of the bestselling science titles of all time.

If you want the brief sense of satisfaction and wisdom that comes from feeling like you finally understand the big bang, the uncertainty principle, black holes, and the quest for a grand unified theory of physics, this is a great read. Just beware the hangover that comes the next day when you try to explain one of these concepts to someone and realise how fleeting and unclear your comprehension really was!

---**THE SPEED READ**---

One of the world's most renowned physicists sets out to explain, with childlike wonder and a helping of good humour, everything about space, time, and the universe. From the tiniest particle to the vastest stretches of space, Hawking gives a brief overview of what we know and what we might know in the future that can make even the least scientific reader feel momentarily like a genius.

Flow:
The Psychology of
Optimal Experience

Mihaly Csikszentmihalyi, 1990

Mihaly Csikszentmihalyi is a Hungarian-American psychologist best known for his contributions to the field of positive psychology, the field focusing on self-acceptance, well-being, quality of life, contentment, and living a meaningful existence.

The concept of 'flow' is essentially living a good life: this doesn't necessarily mean achieving happiness, in spite of the more commercial, but inaccurate subtitle imposed on later editions of the book: 'The Psychology of Happiness'. Csikszentmihalyi is careful to distinguish pleasure that you might find in distractions and sensory experiences, from enjoyment that can only be found through meaningful engagement with an activity. Seeking pleasure rather than true enjoyment is depicted as an inherently doomed task.

In his research, the author asked people to note their state of mind at random moments of the day. The best moments came when people were engaged in an activity they enjoyed. In a diagram, the difficulty of a task appears on one axis, while the skills of the person doing it are on the other. A task that is too easy induces boredom or apathy, while a task that is too difficult provokes anxiety and stress. The happy medium is people creatively engaged in something that challenges them sufficiently to be fully absorbed. It is in this zone that they experience a state of flow: this is what an athlete or dancer would call being 'in the zone', whereas mystics and artists talk of 'rapture' and 'ecstasy'.

Csikszentmihalyi writes, 'It is when we act freely, for the sake of the action itself rather than for ulterior motives, that we learn to become more than what we were. When we choose a goal and invest ourselves in it to the limits of concentration, whatever we do will be enjoyable. And once we have tasted this joy, we will redouble our efforts to taste it again. This is the way the self grows.'

Finding this 'goldilocks zone' (as in the fairy tale, not too much or too little, but just right) is, of course, not easy. But it is not only accessible to yogis and masters of their craft: it can be achieved in all aspects of our life if we know how. We have to identify our goals, find ways to measure progress, and learn to focus on the tasks we undertake. Then we need to keep developing our skills and raise the stakes to keep being challenged sufficiently to maintain the state of flow.

The state of flow is dependent on several conditions: the task must be doable, the target clear, and the person should be able to focus and to receive feedback on their achievements. They also need to be without worry, in control, and to temporarily lose their sense of self and time. Of course, this is a high bar to set, but the important thing is that the closer you can come to a pure state of flow the better your experience will be.

It's also important to note that the experience of flow is

'autotelic', meaning it is an end in itself, rather than something that you do in the hope of reward. As the author says, ' . . . success, like happiness, cannot be pursued; it must ensue . . . as the unintended side-effect of one's personal dedication to a course greater than oneself.'

The author applies the concept of flow to many parts of life, stressing for instance the importance of lifelong learning and exploring what it means to have a purpose in various areas of everyday life. When it comes to the 'meaning of life', he has some fascinating observations. For instance, he effectively says that you can create your own meaning of life, and it doesn't matter what it is. This might sound absurd, but the point is that simply having a meaning and purpose can bring greater focus and absorption to your daily life, and this will bring happiness regardless of the end goal.

This is a better book to read if you disregard the revised subtitle in some later editions, which claims it is the 'Classic Work on How to Achieve Happiness', and remember the author's original focus on 'Optimal Experience', a far more accurate summary of what can be achieved. The state of flow you achieve when you avoid anxiety or boredom and instead find that ideal match of skills and challenges is one way of getting to a good life. The book evokes happiness as constant activity and evolution rather than as a final, achievable state you can directly aim at. The great achievement of the work is to turn most ideas of what happiness is on their head and substitute goals that are more worthwhile.

THE SPEED READ

The optimal experiences in life come when you achieve a state of flow; this is when you lose yourself in a task which challenges you but which you are able to do. By defining purpose in different areas of your life and losing yourself in an achievable challenge, you can bring that state of flow into your life on a much wider level.

Bird by Bird:
Some Instructions on Writing and Life

Anne Lamott, 1994

There are many manuals on how to write, of which the best, in pure technical terms, is probably Stephen King's *On Writing*. However, for advice on how to live your life and think about the world, you might do better to turn to Anne Lamott, whose *Bird by Bird* . . . is a charming exploration of what stops people from writing, and how to write well.

First, let's explain that title. She tells a story about her elder brother, aged ten, struggling to complete a report on birds for his schoolwork. He had had the assignment for three months and it was due the next day. He was sitting at the kitchen table staring at a blank page, surrounded by textbooks, in a state of panic and paralysis. 'Then my father sat down beside him, put his arm around my brother's shoulder, and said, "Bird by bird, buddy. Just take it bird by bird."'

The joy of this book is that Lamott treats the act of writing as something essentially human. For instance, she writes: 'Good writing is about telling the truth. We are a species that needs and wants to understand who we are.' And a large part of the book is about the grit and focus needed to overcome that tendency we all have to become so overwhelmed by the size of a task that we find ourselves unable to take the first step in any direction. As an encouragement to simply get going, she points out, 'Perfectionism is the voice of the oppressor, the enemy of the people. It will keep you cramped and insane your whole life, and it is the main obstacle between you and a shitty first draft.'

As a writing coach, she is very clear that focusing on publication is a trap for the ego. The purpose of writing isn't to be published, it is the writing experience itself that is the reward. She says, 'Writing has so much to give, so much to teach, so many surprises.' She compares it to a tea ceremony: you set out thinking that the point of the tea ceremony is the consumption of caffeine, but then you come to realise that the ceremony is the real point of the exercise. In this respect, it is worth reading this book alongside *Flow* (see p.50), which makes a similar point on a wider scale.

Lamott also has specific advice for writing. For instance, she talks about the experience of writer's block – her suggested remedy is writing at the same time each day and, if you can't think what to write about, simply write about some time in your past that you remember. When it comes to subject matter, she emphasises the importance of writing about something you truly care about. 'The purpose of most great writing seems to be to reveal in an ethical light who we are.'

Her advice on plot and development focuses on understanding your characters and knowing what it is they care most about in the world: this will propel the plot. She gives a formula for short stories which is broad enough to be adaptable for other types of writing: ABDCE, standing for Action, Background, Development, Climax, and Ending. The action grabs the reader's attention, and then you can start explaining the background. Develop the story, reach a climax and find a suitable ending point. This kind of rhythm also works for individual chapters or even for the story arc of a novel.

There's plenty more specific advice in the book, but the heart of it is the way Lamott describes overcoming the inertia that can keep you from starting a difficult task and losing yourself in that task, rather than focusing on the outcome. This wise approach is one that can be applied to many other areas of life beyond the writing process: if you don't know where to start, just remember to take it bird by bird.

THE SPEED READ

Writing, like many other tasks in life, can be a daunting prospect. Lamott's father's advice to her brother on just getting going with his school project stuck with her. Writing isn't always fun or easy, but at its best you can find joy losing yourself in the process. This is the ultimate reward for writing, not any recognition you might hope for.

Guns, Germs and Steel:
The Fates of Human Societies

Jared Diamond, 1997

Any attempt to write a history of civilisation as a whole is a complex business, and here the task is undertaken by Jared Diamond, a professor of geography and physiology at UCLA, and thus not technically a historian. That is one of the reasons why this is such a fascinating read: Diamond's particular interests lead him to some very particular conclusions about the reasons why civilisations have developed in the way they have.

Diamond was once studying the birds of New Guinea when one of the locals asked him a question that stuck with him: 'Why you white men have so much cargo and we New Guineans have so little?' 'Cargo' meant things such as steel, tools and tradable goods in general. One of the traditional explanations given by historians was rooted in racial exceptionalism: the idea that there was something particular in the Eurasian and north African races of the past that allowed them to develop cultures and to conquer other areas of the world.

Diamond gives a very different answer, arguing that geographical and environmental factors made crucial differences. There were certain areas that had a head start when it came to the factors that allowed agriculture to easily develop. This created different types of society in which written language and technology were more likely to flourish, while more dense living conditions encouraged both the spread of germs and increased immunity. Meanwhile, settled societies were more prone to seeking to conquer other settlements, and this led to the development of weapons and war machines of all sorts.

Agriculture first developed in the 'fertile crescent' of the near east and north Africa. There are some very specific biological reasons for this. Firstly, consider the species of the area: wild wheat species such as emmer had heavy seeds that tended to remain attached to their stalks rather than falling to the ground. This made it easy for hunter-gatherers to pick the nutritious seeds and eat them, and led them to discover that the seeds could also be planted to encourage new growth. Several wild pulses similarly lingered in the pod and were easy to harvest. In addition, there were edible wild species – from olives and almonds to chickpeas and figs. And, crucially, there was a range of wild animals that were relatively easy to domesticate – cows, sheep, goats and pigs were all native. This wasn't true of the rest of the continent of Africa, which had fewer concentrated areas of plant species that could easily be harvested and fewer animals that could be tamed. And while agriculture developed in smaller areas in China and Mexico, places with food that naturally offered itself up as a snack and as a seed (for instance, rice, maize, beans, and soya), there were other geographical factors that limited the spread of those civilisations.

The geographic orientation of Europe and Asia was a crucial factor. The Americas allowed for north to south travelling and trading routes, while China was relatively isolated by mountains, deserts and other difficult terrain. By contrast, the geography of Europe and the near east allowed travelling, trading and conquest on an east-to-west axis, in which the variation in climate was less extreme. This meant that the same plants and animals were relatively easy to transplant and transport into different areas of the continent, and it was possible for larger empires to develop and for the agricultural revolution to spread to a much larger area. It was also then feasible for the cultures that developed to make conquests further south into Africa and further to the east into Asia. Meanwhile, the sheer size of the Eurasian land mass allowed for a much greater population growth and, as a result, a larger surge in technology.

Diamond talks extensively about the kinds of positive feedback loops that amplified and exaggerated these differences between the development of the world's civilisations. For example, some diseases – such as measles – flourish in crowded conditions. They act quickly and spread and, while many people die, natural immunity arises in a way that it never will in a hunter-gatherer group who are more likely to be wiped out when they encounter the disease.

As Yuval Noah Harari in *Sapiens* (see p.203) acknowledges, Diamond says that the agricultural revolution did not make life better for ordinary people. Indeed, it could only keep advancing through the growth of highly hierarchical societies, meaning that the have-nots often experienced worse living conditions than any of the hunter-gatherers that came before them. And here is another feedback loop: the greater special-isation that comes with a hierarchical society leads to intelligent people being given more time and resources with which to develop technology. The larger society also tends to foster more of a group mentality, allowing for the development of conquering armies.

In terms of smart thinking, there are a couple of points of interest in this book. One is Diamond's use of the 'Anna Karenina principle' (taken from Tolstoy's book of the same name), which states that if there is a problem with any one of a number of factors in a wider endeavour, then it is doomed to failure. For instance, the development of civilisation in the fertile crescent depended not only on climate and environment, but also on which flora and fauna were available, the existence of certain mineral and metal deposits, sufficient water supplies, and so on.

More importantly, what this book shows is the importance of reframing a question. Instead of asking 'What made Eurasian culture superior?', Diamond asked what geographical and biological factors might have created the conditions for a civilisation to thrive. While some academics quibble with the details of his answer (as will usually happen with a book of this scope), most experts agree that this is at the very least a fascinating way of taking a different perspective on the development of human cultures.

THE SPEED READ

To explain the historic dominance of the civilisations that developed in Europe and the near east, we need to look at the conditions in which they started. Given a propitious set of plants and food, agriculture developed. Given the opportunity to export this culture to a large area with a similar climate, large hierarchical societies developed. In turn, this encouraged the development of technology, including military applications, which enabled the subsequent conquest of large parts of the rest of the world. And it's all because of those germs, seeds and metal resources that were already in the region.

How the Mind Works

Steven Pinker, 1997

Steven Pinker is a professor at the department of psychology at Harvard and specialises in technical psychological areas such as the computational theory of mind. But this is more than an academic tome – Pinker is funny and broad in his interests and takes in everything from silly questions such as, 'Why do fools fall in love?' to the meaning of life, and considers what optical illusions reveal about our psyches, as well as why we lose our tempers. He is a funny, charming writer who keeps the reader entertained while conveying some serious, deep ideas about how and why the mind is the way it is.

Pinker is an advocate of evolutionary psychology, a controversial idea that proposes the human mind has evolved as a device for ensuring our survival and reproductive futures. Most mental traits, especially those that are shared across all cultures, can be explained on strict Darwinian lines. In a sense, this is quite a mechanistic approach to psychology that treats the mind as a powerful computation system.

He is careful to set out some of the basics of this view, exploring, for instance, visual perception, the way that the brain forms three-dimensional images from two-dimensional images in the retina, and the way we then recognise what that image represents in the real world. This leads to a discussion of how neural networks behave, and the way that human thoughts arise from large networks of interconnected neurons.

From this, Pinker moves on through an exploration of many other aspects of the way we think, and our characters and emotions. He focuses on the idea that most of what we experience as intelligence, personality and thought is inherited and can be explained as a product of evolution. Along the way he describes many interesting experiments, such as one in which babies are shown objects and their reactions are recorded. Since babies have not had a chance to be affected by culture, they show the reactions that are hardwired into the human brain.

There are numerous ways in which this kind of theory is frequently opposed. For instance, one traditional problem in evolutionary theories of any sort is the difficulty of designing experiments by which they can be proven. There are also fierce philosophical debates about consciousness. When we experience the colour red, it is because certain wavelengths have stimulated a particular network of neurons. But this doesn't explain how or why we consciously have the experience of seeing something as red. Some writers, such as Daniel Dennett, put this down to linguistic confusion, arguing that there is no need to include the 'I' that is doing the experiencing. Pinker has an alternative view, which is that it is not possible to dismiss the problem of consciousness this easily. He argues that humans are just not smart enough to truly comprehend all phenomena. A hyperintelligent alien who could perceive in more than three dimensions might be able to describe the process, but we wouldn't understand the answers.

Most readers naturally find this a bit frustrating. But there are other more political and religious objections to the whole idea of evolutionary biology. Some people find the whole concept to be dehumanising, suggesting that it removes the 'soul' and undermines the idea of free will. And Pinker has also been attacked by those who fear the field gives support to right-wing thinking on culture and sexuality (such as the belief that some types of human behaviour are simply unnatural), and by feminists who are uncomfortable with the idea that feminine behaviour might be 'hardwired' into the brain. Pinker has dismissed the latter objection, arguing that it is preposterous to suggest that because he believes that 'the male desire for multiple sexual partners has an evolutionary explanation (as opposed to a cultural explanation), [he is] excusing or apologising for men who sleep around.'

Pinker is keen to dispel the idea that his theory is dehumanising and can be quite poetic on the subject of what truly makes us human, arguing that it is crucial also to see people in terms of how widely they draw the circle of people who they treat as being worthy of equal treatment. He argues that seeing the brain in mechanistic terms doesn't undermine the importance of humanity. 'Ultimately, the question is: "How great is the circle?" Does it include the guys in the next village, the guys over the mountain range, children, foetuses, patients in a vegetative state, animals etc? I think a lot of moral debates are not over what is the basis of justice, but who gets a ticket to play in the game.' His aim is not to dismiss the importance of justice and equality, but to understand what underpins the wider way we think about everything in the world around us. Whether or not you accept his argument may depend on how you feel about using evolution to explain mental processes, but he is never less than interesting and this book is a tour de force.

---THE SPEED READ--

A Harvard professor takes a detailed but enjoyable look at evolutionary psychology, the idea that most of our thoughts, character, and behaviour are fundamentally genetic, evolutionary traits. When it comes to understanding how the mind works, this means exploring many aspects of the way we think, react and our emotions in detail.

Irrational Exuberance

Robert J. Shiller, 2000

In 1996, Alan Greenspan was the chair of the US Federal Reserve central bank when he used the term 'irrational exuberance' as part of an attempt to talk the markets down from what he feared might be an unsustainable boom. The idea he was delicately trying to communicate was that the belief of investors in the strength and stability of stock market increases might not have a sound basis.

Robert Shiller's book took its title from the catchphrase, which was in turn partly inspired by Shiller himself. His prediction of a stock market crash was vindicated almost immediately in 2000, the year of the book's publication. In 2005 he added material suggesting that the real estate market was becoming overvalued. It took a bit longer but once again he was shown to be right by the global financial crisis of 2007–08. In the 2009 edition of the book he noted that the boom had 'been ended by an economic crisis of a magnitude not seen since the great depression of the 1930s'.

Among the bulls and bears of the early 2000s Shiller can be seen to stand out as having made some pretty good calls. It's only fair to note that not all of his predictions came to pass, but it's more interesting to examine the thinking that underpinned his description of various markets as overvalued.

He defines the key term thus: 'Irrational exuberance is the psychological basis of a speculative bubble . . . in which news of price increases spurs investor enthusiasm, which spreads by psychological contagion from person to person.' This leads to more stories, in turn used to justify price increases, and more and more investors are drawn in, not because they understand the true value of the relevant asset, but because they are envious of the money other people have made and are spurred on to gamble. Eventually, doubts reach a tipping point and the bubble deflates, although Shiller is careful to note that the deflation isn't truly a bubble 'bursting', since it tends to be a drawn-out process and to proceed in different phases.

There have, of course, been such bubbles throughout modern history, going back to historical crises such as the eighteenth-century South Sea Bubble and seventeenth-century Tulipmania. What is interesting about Shiller's account is his precise delineation of how stocks and other assets can become so mispriced. He points to three common fallacies: prices always tend to bounce back after a fall; stocks will always perform better than other investments over the long run; and investing in mutual funds (a widely trusted form of managed investment scheme) is an optimal plan. Shiller debunks each of these beliefs in turn. For example, if you get into the stock market or real estate at the wrong time then there may be no 'long run' long enough to make up for your initial losses.

Shiller takes a careful look at how closely stock prices are related to fundamental factors such as demand, supply, long-term viability of a business plan and so on. His general conclusion is that pricing is often entirely unconnected to these underlying factors which are actually crucial. When people talk about 'prices' they are often rationalising a situation that cannot be justified.

This is part of Shiller's assault on the 'efficient markets' hypothesis, which suggests that there are many players in a market with different views, and that means the price at any given time will be more or less correct. He shows the many ways in which groupthink and inefficient information can lead to incorrectly priced assets.

An interesting part of the book which demonstrates this is the discussion of new eras. During booms and bubbles there are often commentators who claim that the fundamentals of the market have changed, and that while prices look excessive they have good reasons, since we are now in a 'new era'. An example is the economist Irving Fisher's unfortunate prediction that the stock market had reached a 'permanently high plateau' shortly before the Wall Street crash of 1929 proved him utterly wrong. Shiller points out that if there ever really was a new era, then people would bid up prices at the moment when this became apparent. In fact, the opposite is true: prices rise first, then the argument that this is a new era follows, and this is almost always nothing more than a rationalisation.

The author gives advice on how future bubbles can be restrained, and suggests some good strategies for investors. However, from the point of view of smart thinking, the real interest of the book lies in its forensic uncovering of the many different types of irrationality that afflict commentators, investors and corporations alike. Observing and understanding such epic levels of self-deception is a powerful incentive to always at least consider a contrarian viewpoint. It also shows why you should never stop questioning whether your opinions are based on rational foundations or mere rationalisations.

THE SPEED READ

Participants in the stock market and markets for other widely traded assets are often acting with irrational exuberance, overvaluing assets and rationalising their beliefs. The efficient markets hypothesis is deeply flawed and the many reasons for this are presented in this book. Asset prices often fluctuate without rhyme or reason and it's best to be sceptical of anyone who claims to know the true reasons why this happens.

The Private Life of the Brain

Susan A. Greenfield, 2000

This is an unusual approach to examining the way in which the brain works. As a neuroscientist who specialises in pharmacology, Greenfield has a very physical approach to her subject.

She describes experiments in which animals' brains have been exposed and coated in dye as an investigation into the areas of the brain that respond to light stimulation. She also spends quite a lot of time detailing the different regions of the brain and their functions, as well as digressing into discussion of how long certain drugs take to affect the brain.

Her unusual thesis is that emotions – in the form of our primitive responses and those that develop from our life experiences – are the fundamental building blocks of our minds. This idea is deeply opposed to the more rationality-based view often taken when people describe the functioning of our minds and the creation of our 'identity'. In addition, she is fascinated by the ways in which people attempt to escape from that identity – whether through dancing, sexual release, drug use, or other meditative and euphoric states. She sees this escape as a way out from our more rational selves back to the childish state of pure emotion: in a sense we literally 'lose our minds' at these times. They provide a way to take a break from the more difficult work the brain does when it deals with meaning, logic and memory.

In pursuit of this theory, she writes interestingly on subjects such as children's emotions, road rage, depression, schizophrenia and drug addiction. The essential idea, that emotion is the primal driving force of our minds, is placed within an extended discussion of how mind, body and consciousness relate to one another. She suggests future experiments that could be used to test her theory, which essentially posits that the mind is dependent only upon the brain, while consciousness is dependent on the brain along with other primal systems, such as our hormonal systems. These affect our emotions separately.

She is fascinated by the emotions themselves, and makes some interesting points about the relationship between fear and pleasure (the two sometimes only distinguished by the presence of dopamine), why pain and depression often come together, whether or not fear and pain are mutually exclusive, and what is actually happening in your brain when you are meditating or laughing.

This spectrum that runs from emotion to reason is, of course, not in itself an unusual one. Greenfield makes a case for emotion being a prime mover and has a willingness to explore the reasons why we might want to flee back into pure emotion. This leads to some interestingly different angles on our ways of thought.

She also talks about how our identity develops through our lives. 'We are not fixed entities, certainly not as we grow up, but neither once we live as adults from one year to the next, and even from one day to the next. Even within a day, within an hour, we are different. All the time, experiences leave their mark and in turn determine how we interpret new experiences.'

She goes on to argue that self-consciousness is a fluid thing that ebbs and flows in reaction to emotions, evolving from one moment to the next. Of course, the pursuit of pleasure or oblivion cannot be our constant state of mind or guide to action. And nor would we want to live in a constant state of an intense emotion, such as fear or anger, even if those sensations return us to a childlike state.

Part of the challenge we face in life is how to reconcile emotion with a more rational sense of self. Seeking only to revert constantly to pure emotion would be self-destructive, something which Greenfield memorably describes as 'the paradox of adult human existence'.

Greenfield does digress from her main subject into pedantic asides and, as so often with academics, you occasionally get the feeling she might be settling scores in obscure disputes we aren't aware of. But this is an interesting and knowledgeable overview of how the brain works and thought provoking in its thesis.

---**THE SPEED READ**------------------------

The most fundamental parts of our minds are our emotions, which are rooted in our brain and our primal systems, as opposed to our rational mind which is purely rooted in our brain. We spend a proportion of our time trying to escape our rational minds and 'identity' through the release of sex, drugs and rock 'n' roll, partly in order to return to a childlike state in which emotion is once again the dominant force.

Fooled by Randomness

Nassim Nicholas Taleb, 2001

Nassim Nicholas Taleb is probably best known for his 2007 book *The Black Swan*, in which he discusses the role of highly improbable events, including major disasters and bestsellers. These events are difficult to predict and might occur at any time, yet we attempt to rationalise them after the event and to blame people for not predicting them, so they are a good example of how our thinking tends to be flawed.

Fooled by Randomness is his earlier work on the wider subject of chance in our lives. He discusses luck, probability, uncertainty and how decision-making based on those factors can be afflicted by irrational thinking and error.

Let's get the elephant in the room out of the way first. Taleb is an egotistical writer, and if you are bothered by grand-standing, boasting or being patronised, you might find his style rather overbearing and might prefer someone a bit more emollient.

On the other hand, Taleb is genuinely smart and widely read, and has many fascinating things to say, and you might want to accept that he can be pompous, see it as amusing and just be glad you haven't been seated next to him at a dinner party . . .

Anyhow, the overriding theme of *Fooled by Randomness* is that chance plays a far larger role in our lives than we are willing to admit. We naturally look for patterns and reasons and our brains don't deal well with randomness; we often rationalise events that occur for entirely random reasons.

Another major part of Taleb's theme is the idea that life is non-linear and there are tipping points in which a certain piece of action will have a disproportionately large effect. An example is the use of the QWERTY keyboard, originally introduced to make it harder to type quickly so that typewriters wouldn't jam. If typewriter design had been improved in those early days, the keyboard might have been made to behave more efficiently. As it was, at some time we reached a tipping point – one more QWERTY user meant the disadvantages of so many having to learn a new way of typing outweighed the potential advantages.

Taleb makes many of his points using anecdotes and investor characters he names Nero and John. Nero is a sensible, conservative investor. John takes huge risks but hits a string of winners, has a higher social status and a better reputation as a result. Taleb argues that this tendency to base our judgment of people on the results they obtain rather than on the probability of success is as irrational as acclaiming someone for being brilliant at Russian roulette if they happen to survive a few games with one bullet in the barrel. He advises us to judge people as heroes or villains based on the actions they take, not on the outcomes, especially where the outcomes are dependent on probability.

Taking a longer view of history, and of stock markets in particular, is one way to acquire greater data and to gain a clearer understanding of the role of chance. This allows us to focus on the randomness that almost always underpins action, rather than making the mistake of imposing patterns on data from shorter periods of time.

Taleb has some interesting asides to this train of thought, pointing out that even evolution can be fooled by randomness if, for instance, a tree falls in the woods and kills the best adapted member of a particular population. Similarly, a great company might show poor results purely because of random conditions.

He also talks about the idea of bull and bear markets that seize control of the imaginations of investors and lead to some irrational behaviour, in spite of the fact that over a longer period it is very unusual for there to be a significant difference in the balance of days when the market rises and falls.

He also discusses widely debated cognitive biases. For instance, the whole point of a black swan event is that we use induction (the logical process by which we predict future events based on the past) to predict that all swans are white when we have not visited Australia. We are then taken aback by the results we find there – that the black swan can be found in that country.

And he criticises what he sees as survivorship bias in books such as Thomas J. Stanley's *The Millionaire Next Door*, which, he argues, focuses on individuals who have been lucky enough to have continued success. This ignores the data from other investors who have been wiped out making similar decisions. Take, for instance, a population in which half of the participants randomly beat the market in any given period. After two periods, a quarter of the participants will have been winners twice. After five periods, you will be down to about 3 per cent of 'survivors' who have been lucky enough to beat the market each time. Focus your investigation into how to be a winner on these individuals and you will certainly be letting yourself be fooled by randomness. There are many other fascinating insights into the consequences of such cognitive biases in the real world throughout the book.

Taleb's main conclusions are that we always need to bear in mind the role of randomness. However, our emotions are irrational and we often overreact. Sometimes we need to do so in order to make a decision, particularly in a situation in which there is no optimal outcome. This is highlighted, for instance, in the thought experiment of 'Buridan's ass', when a donkey that usually goes to the nearest food source is presented with an equidistant portion of food and water, and becomes frozen in indecision as a result.

We need to take pleasure in randomness – as long as it is not having a harmful effect on us – and to develop a stoic attitude so that we can deal with those occasions on which random events work against us. And we always need to remember to value people (including ourselves) on the choices they make rather than on the random outcomes they provoke.

THE SPEED READ

We underestimate the role of randomness in our lives at our own peril. Don't believe success stories, as they are often just the story of people who were lucky. Bear in mind that good fortune can be temporary and arbitrary. Remember that analysts are often imposing orderly patterns on data that is, seen from the right perspective, random. Don't be overconfident in your own abilities, and develop a stoic attitude. And, above all, never ever forget that Nassim Nicholas Taleb is smarter than the average bear . . .

A Short History of Nearly Everything

Bill Bryson, 2003

At more than five hundred pages this is not an especially short history at all: however, given the astonishing amount of ground it covers it is a miracle of concision. Bryson's work essentially covers the natural sciences, from cosmology to palaeontology, and chemistry to particle physics.

At a certain age, Bryson realised that he found some of the scientific concepts that had seemed so dull at school utterly fascinating. He set out to educate himself by interviewing leading scientists and compressing their insights into a book that tells the story of science and scientists.

Part of the joy of this is the irreverent tone the writer takes in describing the physicists, biologists and zoologists who helped develop our understanding of the world. He is wryly amused by how much randomness, eccentricity and serendipity was involved in the process. For instance, he relishes such stories as that of Henry Cavendish, who was able to calculate the weight of the Earth but was so shy he never got round to telling anyone; of a Norwegian palaeontologist who mis-counted the digits on the hands and feet of a crucial fossil discovery and compounded the error by refusing to let anyone else examine it; and the Victorian naturalist who amused himself by serving mole and spider to his dinner party guests. At the same time, Bryson makes the effort to debunk some popular myths about scientists, including the idea that Darwin made a great leap of understanding on examining the beaks of those Galapagos finches.

The book is also good at conveying the extraordinary way that atoms have combined, separated and recombined over and over since the big bang, leading to some odd observations like the fact that it is remarkably likely that we each contain at least one atom from the body of William Shakespeare. This is the kind of thing they didn't teach in chemistry lessons at school, but that can really focus your attention on the importance of chemical reactions.

The key thing with Bryson is that he doesn't rest until he finds a way to get his scientific sources to explain their subject in the most comprehensible way possible – whether it be the question of how evolution works, what a black hole is or how the continents of the Earth drifted around in the past. He finds a clear, simple way to explain it in layperson's terms. Bill Bryson has always had a nice way with words, and paints visual, friendly imagery to help the reader – for instance, claiming that 'protons give an atom its identity, electrons its personality'.

Bryson is also good at capturing the wonder of the things that science has discovered about the world. He points out for instance that every living cell has about the same number of working parts as a Boeing 777, and that dragonflies the size of ravens used to fly through the moss-covered forests of the prehistoric period.

In addition, as an essentially humorous writer, he finds the odd nugget of comedy in the travails of early scientists and discoverers, pointing out for instance that the identification of the compound phosphor came about because an alchemically minded scientist, convinced by the similarity in colour, was attempting to purify urine to make gold.

As with any attempt to cover the history of absolutely everything, there are a few gaps and occasional moments when too much seems crammed in too fast. But on the whole the book is a terrific read that will reintroduce the reader to those subjects that seemed so dry at school. And, at all times, Bryson succeeds in conveying his childlike wonder at the complexities of the universe. 'Tune your television to any channel it doesn't receive and about 1 per cent of the dancing static you see is accounted for by this ancient remnant of the big bang. The next time you complain that there is nothing on, remember that you can always watch the birth of the universe.'

─THE SPEED READ─

Science is actually much more interesting than it seemed at school. You just need to see the big picture, from the creation of the universe, through the combination of particles, to the magic of the creation of life and the evolution of humans. Bill Bryson gives a charming, concise account of the science and the often eccentric scientists who first stumbled on the great discoveries.

Moneyball:
The Art of Winning an Unfair Game

Michael Lewis, 2003

Moneyball is a highly influential book about how our assumptions about the most important features of a sports fixture or team can be utterly wrong.

The most important character in the book is Billy Beane, the general manager of the Oakland Athletics baseball team, also known as the Oakland A's. The team was short of money and couldn't afford the kinds of star players that would usually be pursued by a club looking to improve results. Beane's solution was to instruct his staff to carry out extensive statistical analysis of some of the less obvious features of the sport that might influence the outcome of games – for instance, hitters who had a high on-base percentage. Where there was a correlation between such a feature and success, he pursued less expensive players who, nonetheless, displayed the necessary statistical trends and features. The result was highly successful as the Oakland A's massively improved their performance.

The term 'moneyball' is now widely used to refer to teams in any sport who rely on such detailed statistical analysis, which has become much more widespread because of Lewis's book. The author has acknowledged the ironic outcome: as other teams began adopting the same methods, the Oakland A's competitive advantage was hit and their results suffered.

Lewis discusses several other themes. Before the moneyball methods were adopted, many in baseball and other sports relied on traditional methods to gauge players, including stolen bases, runs batted in and batting average. He describes the initial users of sabermetrics (the name for such analysis) as upstarts, disruptors and outsiders, as opposed to insiders who used traditional methods.

He also discusses the idea that we live in an age when information has been democratised: if everyone has access to the same information, then in theory no one has a competitive advantage. The old hierarchies are flattened out and only those who can adapt and find new uses for the data prosper. Given what he describes as the 'ruthless drive for efficiency that capitalism demands', such adaptation is an ongoing process, especially for the underdogs. When other teams adopted sabermetrics, Oakland had to stay ahead of the curve and take into account more complex player attributes, such as defensive capabilities.

Of course, this problem of keeping ahead of the pack existed before the internet. Any kind of system relying on statistical analysis works best before it is widely known, for a variety of reasons. It is well known that players at clubs relying on statistics work hard to improve their performance in a particular area such as that of 'ground covered' in a soccer match, to the detriment of the efficiency in predicting success. And sports evolve over time. In the 1970s, the ability to hit more sixes than average in cricket would have been an amusing, but risky feature of the more reckless players, whereas now that 20:20 cricket has become hugely popular and lucrative, it is an essential skill that all serious batsmen have to work on.

Another good example comes from the 'Beyer speed figure'. This method of rating the performances of racehorses was published in the book *Picking Winners* by Andrew Beyer, a *Washington Post* columnist, in the early 1970s. It allotted 'par speeds' for a particular racetrack for each horse based on past times. Using the Beyer figure gave a gambler a real advantage until handicappers and bookmakers caught on to the method and began to price them in to their odds. The competitive advantage was destroyed.

Anyone relying on statistical analysis, whether it be in gambling, sports or business, needs to be nimble and to accept that what works at one time may later go out of date as the system becomes more widely used or as the sport or business evolves over time. It seems that sometimes, even when you deploy a bit of smart thinking, you have to be smarter still to work out how to keep on making it relevant.

Moneyball has changed sport in many ways and this is a classic book by one of the most interesting and enjoyable writers on business psychology. The concepts it introduced are now so familiar that it almost seems like old hat, but it does retain interesting ideas that are still highly relevant today.

THE SPEED READ

The story of how Billy Beane managed to turn around the fortunes of the Oakland A's baseball team by using statistical analysis (aka sabermetrics) has influenced the entire world of sport into using more inventive ways of analysing performance. Lewis's engaging account of the feat also includes interesting thoughts on insiders and outsiders, and the democratisation of information in the digital world.

Train Your Brain:
60 Days to a Better Brain

Ryuta Kawashima, 2003

The Japanese neuroscientist Ryuta Kawashima is an absolutely fascinating figure: his academic work centres on explorations of the regions of the brain that control various aspects of our cognition.

He's probably the most famous neuroscientist in the world with millions of fans who know him as the animated professor in Nintendo's *Brain Training*. The game, created following the huge success of this book, has sold over twenty million copies. Kawashima reportedly turned down an annual salary of fifteen million euros for the game role, instead drawing a salary of seventy thousand euros and putting the rest of the money into funding research. He says he believes you shouldn't be paid such riches unless you have truly earned them.

So, why was his book so popular? Essentially, his study of brain-training has led him to a simple system in which repetitions of certain daily tasks are used to stimulate the memory. Each day, the user spends a set amount of time (generally just fifteen minutes) carrying out mental calculations, reading aloud and writing. Improvements in mental acuity are tested with Stroop effects – named after their inventor, John Ridley Stroop, they are exercises in which the participant names the colour of a word printed in a different colour. For instance, if you see the word 'blue' written in a green font, you would answer 'green'. The research is based on neuroplasticity, which suggests our neural networks can continually be rewritten and modified as we grow.

These exercises are recommended for everyone from students to the elderly, especially those who are starting to forget things. The BBC journalist Adam Shaw visited Kawashima's 'brain gym' in Tokyo a few years ago and found himself being trounced in a memory contest by one of his students, the octogenarian Endo Tokiko. Brain monitors showed that, while Shaw's brain had been firing on all cylinders, Tokiko's had barely flickered, indicating she was barely trying. Shaw wrote, 'In other words, Kawashima explained to me, not only had I been beaten badly but my opponent had done it with one arm tied behind her back. She had used only a fraction of her brain power, while I had brought everything I had to the game.'

That Kawashima's programme can have such demonstrable results suggests it is an effective method of improving memory. And what of the book? Well, there's not much to describe. Kawashima takes a few pages to outline the neuroscience that lies behind the method he is recommending. The rest of it is tests and answers, along with explanations of how to use them – this is an entirely practical manual, a training method in book form, and one that gets results. And if that isn't enough to sell it to you, you could always go to YouTube and enjoy watching the esteemed doctor's disembodied head floating around in games like *Super Mario Maker – Dr Kawashima's Athletic Training*.

THE SPEED READ

A mental training method, created by a megafamous neuro-scientist who has millions of fans in Japan and around the world. Take fifteen minutes a day to work through these calculations, to read aloud, write and take a few other tests, and your memory and mental acuity will benefit.

Being Logical

D. Q. McInerny, 2004

This is a wonderful, short guide to logic: in about 150 pages it covers all the absolute basics, while cramming in a serious amount of advice about how to spot and avoid illogical thinking.

It starts with a section looking at the basics that underpin any attempt to reason about the world: facts, which are objective things and events; ideas, which are the way we subjectively represent facts in our minds; words, which are the building blocks of language; statements of propositions, which combine words to express a belief or view of the world; and knowledge, which is the ultimate goal of all of the above.

The author then moves on to discussing the basic principles of logic. In particular, he says, we won't get very far using logical thought if we disregard the fundamentals:

- The principle of identity, which means that a thing can only be itself.

- The principle of the excluded middle, which says that a proposition is either true or not true.

- The principle of sufficient reason, which says that everything has a cause or reason.

- The principle of contradiction or non-contradiction, which says that statements that contradict each other can't both be true.

There are philosophical quibbles that could be noted (for instance, a proposition may be defined in such a way that it can be meaningless or not, which would affect the second principle). But these principles, which are self-evident – meaning we needn't prove them – nonetheless underpin all real practical situations in which logic can be applied.

At this point, McInerny takes time to observe that, before you even get to the point of using logic, you need to make sure you have taken time to check your basics. Are the things you are accepting as 'facts' true? Are both parties defining the terms in the same way? Are you mistaking correlation for causation, and thus ascribing the wrong causes to facts? Unless you have got these fundamentals right, you won't succeed in forming a logical argument that produces true results. He writes, 'Bad ideas do not just happen. We are responsible for them. They result from carelessness on our part, when we cease to pay sufficient attention to the relational quality of ideas, or, worse, are a product of the wilful rejection of objective facts.'

He moves on to defining an argument, which consists of two parts: the premise and the conclusion. Within the overall definition of an argument, we can separate deductive arguments from inductive ones. A successful deductive argument delivers a conclusion that is necessarily true, whereas an inductive one will only deliver a conclusion that is probably true. It's worth noting that even some of our most fundamental beliefs are based on inductive reasoning (for instance, 'The sun will rise tomorrow').

McInerny points out that arguments can be both an art (when their purpose is persuasion) and a science (when their purpose is to find the truth). But he is more concerned with the latter purpose. A good argument generally needs to fulfil particular requirements: the premises need to be true and relevant, the logic needs to be valid and, when it comes to inductive arguments, the premises should be strong enough to support the conclusion.

It's worth noting that occasionally McInerny makes sweeping statements that a formal logic geek would question. For instance, he overstates the first of the requirements above when he writes, 'If we start with a false premise, a valid (i.e. structurally sound) argument will only allow us to proceed consistently to a false conclusion. The adage "garbage in, garbage out" applies nicely here.' This is demonstrably wrong since false premises can support true conclusions, as in the argument 'Sven is in Sweden, Sweden is in Scandinavia, so Sven is in Scandinavia', if it is made when Sven is actually in Finland. The country may be wrong, but both Finland and Sweden are in the region of Scandinavia. However, McInerny's statement is good enough when it comes to giving advice on practical logic, where it is, of course, best to rely on true premises.

Finally, in the most indispensable part of the book, McInerny discusses the 'sources of illogical thinking'. In particular he looks at twenty-eight of the most common logical fallacies and ways that arguments can be distorted. These can be used not only to make better logical arguments, but also to see when other people are using illogical ones. For instance he discusses 'reductionism', in which a complex problem is reduced to oversimplification; the 'straw man' argument, in which a proposition is intentionally put forward merely because it is easy to reject; the ad hominem argument, which depends on undermining the person making the argument; the false dilemma, in which a wide variety of options are artificially narrowed down to two, one of which is usually clearly unfavourable; special pleading, in which people omit information which might undermine their own argument; and self-explanatory strategies such as 'begging the question', red herrings and the use of laughter (or tears) as a diversionary tactic.

This is an extremely useful, brief manual on the art and science of argument, enhanced by the positive, upbeat approach of the author, who writes, 'In the ideal debate, the primary purpose of the debaters is not to triumph over each other, but rather by their combined efforts to ferret out the truth as it pertains to the issues being debated.'

If only this motto were to be adopted by politicians, the media, and online commenters, the world would be a much, much better place.

THE SPEED READ

Logical thinking is the best way to establish the truth about the world and to make good decisions. It is important to understand the absolute fundamentals of what arguments consist of, what distinguishes a good argument from a bad argument, and when you should and shouldn't trust the argument someone else is making.

Mind Hacks:
Tips and Tricks for Using Your Brain

Tom Stafford and Matt Webb, 2004

As part of publisher O'Reilly's 'Hacks' series on a range of subjects, *Mind Hacks* has a format and structure that feels closer to a computer-training manual than the average book, which might make it seem like it is targeting a geekish audience.

This feel is occasionally emphasised in the titles of the hacks (for instance 'Neural noise isn't a bug, it's a feature'). But it's worth looking beyond the surface because this is a fascinating book that shows you ways you can start understanding some of the things that are happening, without your knowledge or permission, in your brain on a daily basis.

The book opens with a neat analogy to describe cognitive psychology, the approach used in the work, in terms of reverse engineering. The authors explain that you have to study the brain without being able to see inside it, since it is in many respects an impenetrable 'black box'. They compare this process to trying to work out how the Google search algorithm works. We might, for instance, wonder whether word order makes a difference to how Google responds. You can try this experiment yourself. Do a Google search first for 'reverse engineering' and then for 'engineering reverse' – you get different responses. You may conclude that there is something in the Google algorithm that does respond to word order, even if we can't read the exact line of code. Neat, huh?

The 'hacks' in the book are basically ways in which we can 'see inside the brain', including EEG (electroencephalogram) readouts and unusual treatments like transcranial magnetic stimulation, which is used to treat depression by 'turning bits of the brain on and off'. The most interesting hacks are those that encourage the reader to carry out tiny experiments on themselves that have been used on a wider audience. The types of cognitive processes involved in this include vision, attention, motor skills and subliminal perception. These are generally given an entertaining and vivid explanation.

For instance, the authors discuss the phenomenon of 'negative priming' by giving a witty summary of the story of the boy who cried wolf. As a result of raising a false alarm, he ended up being ignored by the villagers (and ended up as wolf food). In a similar way, the brain responds more slowly to a stimulus that has previously been used to mislead it. The example given in the book is one you can try yourself: a flashcard is printed with a prominent black star and a greyer star. You are asked to name the grey image and with that your brain learns to discard the 'star' information. The next flashcard has a black cloud image, with a fainter grey star. When asked to name the grey image, there is a recordable difference in the time taken by subjects to do so compared with the time they took with the first card. This is because the brain's first instinct is now to ignore the 'star' information. A similar phenomenon has been recorded when subjects are asked to pick a red pen from a cup containing a red pen and a blue pen, before being asked to pick the blue pen.

Another fascinating test that you could, in theory, do at home (although it takes months) is based on an experiment first carried out by the scientist Vinoth Ranganathan. Two groups were asked to train five days a week for twelve weeks, putting in fifteen minutes' 'exercise' a day. The first group were told to place their palm flat on the desk and to extend their little finger so as to push an object outwards, thus strengthening the muscles involved in this unusual movement. The second group were only asked to imagine tensing the muscle in this way.

Unsurprisingly, the first group's ability to push an object in this way improved: they were able to exert 53 per cent more force at the end of the experiment. More surprisingly, the second group were able to exert 35 per cent more force, meaning that they had become stronger merely by using their imagination.

There are a hundred hacks in the book, and a huge number of small revelations along the way. We're introduced to the way that our eyes are actually in constant motion, but our brain tunes this movement out; we're shown how to detect sounds that are right on the edge of audibility; we're taught to boost our memory through a focus on context and much more; and we're given a fascinating tour of hypnagogic states (the dreamlike moment between wakefulness and sleep).

In addition, we can pick up many fascinating nuggets of information. For instance, did you know that you actually live 80 milliseconds in the past – the length of time it takes your brain to process perceptions. And, did you know that, in general, when people read they have a tendancy to take in only the first and last letters of a word . . . which is why you might not have noticed the misspelling of 'tendency' earlier in this sentence.

---THE SPEED READ---

Why not do some experiments on your own brain? Don't worry though, you don't need to start sticking probes into your skull: all you need to start seeing inside the 'black box' of your brain is the kinds of experiments that are carried out by cognitive psychologists. They can give us some very clear information about what our brain is doing, even if they don't always know the exact reasons for the behaviour.

Blink:
The Power of Thinking
Without Thinking

Malcolm Gladwell, 2005

O h, those one-word titles . . . those snappy, counter-intuitive subtitles . . . publishing marketing departments sure do have a lot to answer for.

Malcolm Gladwell has built a career on books that make pretty bold, unexpected claims. In this case, he is suggesting that decisions made on the hoof, in the blink of an eye, can often be as good as or better than ones made after due consideration.

He bases this claim on our skill at 'thin-slicing', the ability to use a small amount of information from a limited time period to reach a sound conclusion. He gives examples from all kinds of fields including science, sales and marketing, dating and romance, and culture. For instance, he tells the story of the Getty kouros, a statue brought from the Acropolis in Greece to an American museum. George Despinis, head of the Acropolis Museum, reportedly dismissed its authenticity on first sight, saying, 'Anyone who has ever seen a sculpture coming out of the ground could tell that that thing has never been in the ground.' He was proven correct.

In a similar vein, Gladwell talks about John Gottman, an expert on relationships, who can predict whether or not a couple will divorce with 90 per cent accuracy after only fifteen minutes of observation.

Gladwell's thesis is that too much information can get in the way of good decision-making, while our unconscious is actually a pretty good information filter. Given the sheer amount of information we can access, experts can get bogged down in a detail quagmire, while their gut instincts are telling them the right answer. Rather than getting stuck in (rhymey jargon alert!) 'analysis paralysis', a snap judgment may be the best resolution.

Of course, the idea that spontaneous decisions are always superior is a pretty controversial one, and Gladwell is forced to confront the fact that it clearly isn't universally true. He discusses the shooting of Amadou Diallo, an innocent man killed on his New York doorstep in the US by cops who made a bad, intuitive decision. Gladwell is also forced to acknowledge the obvious fact that it takes a lot of experience, learning and practise to get to the point where you are expert enough to make good spontaneous decisions.

In order to explain why his big idea doesn't work, he introduces a range of prevarications. He talks, for instance, about the way in which your unconscious biases and prejudices can get in the way of a good decision. He also discusses implicit association tests that can reveal such unconscious biases and psychological priming. Our experiences affect later perceptions and decisions.

Of course, this is all pretty obvious: the whole point of taking time to make a decision is to try to avoid such cognitive biases. But Gladwell has a few other smokescreens to deploy. For instance, he goes on to consider the role of stress on a decision. In the same way that autistic people are sometimes unable to pick up on nonverbal clues, stress can temporarily rob us of our ability to pick up on the kinds of microclues that our subconscious is generally processing. To deal with these sorts of issues, he suggests putting up mental or physical screens that cut out the factors that might lead you into wrong decisions. For instance, a record company scout may be distracted by a singer's appearance, in which case it might make sense for them to review audio demos only.

This is a case where it is worth listening to some expert opinions on Gladwell's thesis. The US appeals judge and professor Richard Posner wrote, 'Rarely have such bold claims been advanced on the basis of such flimsy evidence.' The author Michael LeGault (who published a book called *Think* as a rebuttal of *Blink*) has argued that '*Blink*-like' decisions can be dangerous while fiercely disputing Gladwell's claims that 'mind-reading' is something that only goes wrong 'sometimes'. The distinguished author and psychologist Daniel Kahneman has written that Gladwell doesn't really believe his own thesis 'that intuition is magic. He really doesn't . . . But here his story has helped people, in a belief that they want to have, which is that intuition works magically; and that belief, is false.'

There are, inevitably, interesting anecdotes and examples in the book: that is one of Gladwell's undeniable strengths. But the underlying argument is so flimsy that the author has to spend most of the book justifying it against all the entirely obvious objections. This suggests that this was one contrarian idea that should have stayed on the shelf.

┌─ **THE SPEED READ** ─────────────────────────┐

People make spontaneous decisions that are as good as or better than those reached through critical thinking and reflection. OK, that's obviously not always true, but that's just because of cognitive biases . . . or not having enough experience . . . or because of emotions and other stuff. Quick, look at the shiny thing over there . . .

└──┘

Freakonomics:
A Rogue Economist Explores the Hidden Side of Everything

Steven D. Levitt and Stephen J. Dubner, 2005

Freakonomics is a cultural phenomenon, merging pop culture with economics, making the latter far more fun and accessible. It is also a terrifically enjoyable read. Its snappy chapter titles like, 'What do school teachers and Sumo wrestlers have in common?' and 'Why do drug dealers still live with their moms?' are immediately eye-catching and draw the reader into some quite dense economic explorations. Levitt, in particular, was already well-known for applying economic theory to unusual areas of life, and mining reams of statistical data to come up with often counter-intuitive theories.

The book focuses strongly on the difference between causation and correlation, and on situations in which people may have confused the two. Each chapter takes a quirky subject: cheating and why people cheat (with reference to those teachers and wrestlers, as well as to a bagel business); information control (with reference to the Ku Klux Klan); the economics of drug dealing; the role legalised abortion has played in reducing crime (with reference to the regime of Romanian dictator Nicolae

Ceauşescu; the lack of correlation between good parenting and education; and nominative determinism (the idea that the name you give a child will have an impact on its future life).

One example of the book's ingenious use of statistics is its claim to have proven that sumo wrestlers cheat. All wrestlers compete in fifteen matches in a tournament, and will be demoted if they don't win more than half. The authors compare results in the final match, in particular where a wrestler with seven wins and seven losses is opposed by a fighter who has already won eight matches. In theory, the 8–6 wrestlers should have a higher chance than the 7–7 wrestler, since they are statistically slightly better performers. However, the 7–7 wrestlers win as much as 80 per cent of the time. This suggests, according to the authors, collusion in the tight-knit Sumo community, as the fighters with eight wins are reluctant to knock out their opponent. The Japan Sumo Association said they were investigating, and the 2011 Grand Tournament was cancelled due to match-fixing allegations.

When it comes to the material on drug dealing, the authors are at pains to point out they are not making moral judgments about the topic (in technical terms, they are engaged in positive econometrics, not normative economics). They analyse the financial records of the Black Disciples, a gang in Chicago, and found that many dealers were earning remarkably small amounts – as little as three dollars an hour. They use this to discuss the concept of a winner-takes-all labour market, in which many workers are competing for a slot in the marketplace, but few are succeeding, while the 'winners' are getting paid much higher salaries. In the Chicago gang, more than half of the profits were earned by about 2 per cent of the gang members. They refer to this kind of market as a 'tournament', in which there are clear winners and many losers.

There is much else in the book that is revealing and surprising, including some fascinating material giving insights into the experiences of people who have infiltrated the Ku Klux Klan and a criminal gang. However, as the book progresses, some of the material becomes a bit weaker and some fairly obvious points are missed by the authors. For instance, they claim that, as the very safe country Switzerland has more firearms than most others, it can be proven that guns don't cause violence. However, it could equally be explained by the extremely tough laws on ammunition in the country which are equivalent to fairly tight gun control by the government.

There have also been some fairly academic, but legitimate objections to the authors' uses of statistics, along with the suggestion that their pop-culture approach has led them to blur some complex issues in an unhelpful way. For instance, one reader notes that in the section on fatal crashes they have relied heavily on the database known as the Fatality Analysis Reporting System and they rely on identification of uninjured children in crashes that were fatal for other passengers. However, the historic data set from before 2002 is not as complete or consistent as they seem to assume. So it is claimed that some of the counter-intuitive results that overturn the received wisdom rest on less reliable grounds than they might appear to.

The book caused a more serious controversy with its claim that violent crime fell after the abortion ban was overturned. The economists Christopher Foote and Christopher Goetz published a paper arguing that there were both statistical errors and significant omissions in the data set that the authors' argument rested on. They pointed out this didn't necessarily contradict the conclusion reached, but argued that when the data is incomplete and messy it can be hard to prove a theory either way.

The Economist responded to the paper by commenting that 'for someone of Mr Levitt's iconoclasm and ingenuity, technical ineptitude is a much graver charge than moral turpitude. To be politically incorrect is one thing; to be simply incorrect quite another.' The authors corrected some errors and argued that the paper by Foote and Goetz was inaccurate in other respects, but the debate was sufficient to raise some doubts among academics about the book.

Freakonomics is not a book that comes without baggage and problems, and cautious readers may want to fact-check any particular claim made in it (or at least to check for later rebuttals and controversies). But it is nonetheless a fascinating exploration of the ways that data can be used to reveal the true patterns that underpin our everyday lives and is, at the very least, a rewarding and amusing trip through the world of thought experiments and statistics.

THE SPEED READ

A whirlwind tour through the worlds of drug dealing, parenting, education, cheating, informational imbalances and Sumo wrestling, taking a pop-culture hammer to the vase of received economic wisdom. Great fun and occasionally revelatory, but also a book that is worth reading with a sceptical mind.

Stumbling on Happiness

Daniel Gilbert, 2006

You might get a sense of whether or not you will warm to Daniel Gilbert or not from the way he starts chapter one: he muses on the fact that every psychologist has taken a vow to write an article or book that includes a sentence starting 'The human being is the only animal that . . . ' He points out the inherent danger in this, noting that if you had chosen to end it with 'can use language' or 'uses tools' then you are setting yourself up to be humiliated by a monkey as soon as someone discovers a chimp using hand signals or a stick to fish termites out of a mound.

This is typical of the author's dry, humorous tone. He risks his own entry in the list of similar sentences when he suggests, 'The human being is the only animal that thinks about the future.' And that is essentially what this book is about: our ability to imagine the future and how happy it will make us. The overall suggestion is that we are pretty bad at imagining our future for a variety of reasons, and that when we imagine bad things happening in the future, we often overrate how much pain they will cause us.

Gilbert is especially interested in the ways in which our brain misleads us. He talks about the blind spot, a patch of cells in our retina which do not register light. This means that in theory there is always a black hole, a gap in the information we perceive. However, the brain fills in the gap with a reasonable estimate of what might be there. If, for instance, there is a clear pattern then our brain will fill the gap with that pattern. What this demonstrates is that, whether we want it to or not, the brain invents part of our reality.

When we picture concepts such as memories, the future, and our happiness levels, there are a variety of ways in which this invention goes wrong. Our imagination does something similar to the brain, manufacturing details that aren't actually present. We don't realise it is doing this and accept the image at face value.

We are also too stuck in our present and will interpret both our memories and the future through that filter: we imagine the past and future as more like the present than is the case. And finally, we have a 'psychological immune system' that ensures events do not seem to be as bad as they did when we contemplated them. For example, when people are tested for a genetic problem, their happiness obviously rises if they receive the all-clear, yet the feeling is also increased by a positive result: once the bad news has been received, the brain intervenes to make it seem not so grim. The only situation in which there is no change is if the patient is given an inconclusive result.

The book is divided into six sections. The first, Prospection, deals with that tendency to 'fill in' details of predictions or memories. The second, Subjectivity, discusses the difficulty of measuring happiness: 'evaluating people's claims about their own happiness is an exceptionally thorny business'. The only way to understand trends is by comparing first-hand responses to questions over large groups of people. Over the following three sections, Gilbert explores some more of the problems faced by our imaginations: the lack of realism, a bias towards expecting the future to be like the present, and the way in which we alter our own perception. 'A healthy psychological immune system strikes a balance that allows us to feel good enough to cope with our situation but bad enough to do something about it.'

In a book like this, one might expect some advice on how to overcome cognitive biases. Gilbert does suggest one way we can do this, but glumly predicts that we won't take his advice. His advice is that we will get a better prediction of our future if we compare ourselves with other people: we are quite likely to have similar experiences to those who have been through the same things. The problem is that we find it hard to see ourselves in this way, as a single data point in a crowd of similar individuals. As he says, 'If you are like most people, then like most people, you don't know you're like most people.' Very droll.

Don't come to this book for self-help advice. And if you find a witty but somewhat self-satisfied tone irritating, you also might want to skip it. But Gilbert can be genuinely amusing and there is a lot of interesting content here that will affect the way you think about your own memory and imagination.

THE SPEED READ

Our brain fills in gaps in our perceptions and memories, making our imagination unreliable as a tool for predicting our future. Events usually turn out to be better than we expected them to be, if only because our brain compensates for bad experiences in a way that we find hard to imagine in advance. If you want to predict your future, abandon your feelings of individuality and accept you are just part of a crowd, one that shares many predictable traits.

A Whole New Mind:
Why Right-Brainers Will Rule the Future

Daniel Pink, 2005

When it was published, this was a popular, successful title that essentially argued that creativity is the key to future business, as it is the only thing that can't be outsourced. With that in mind, it's best if we get the elephant in the room out of the way at the start.

Pink bases his entire thesis on the idea that there are such categories as right-brained and left-brained people. Ever since it was shown that the left and right hemispheres of the brain function differently, people have been fascinated by the distinctions and relations between them. The left brain is stronger when it comes to verbal logic, maths and linear thinking, whereas the right brain is associated with creativity, holistic thinking, imagination and visualisation.

The problem is that, since 2005, there has been a study of sufficient size to more or less prove that there are no such categories of people as right-brained and left-brained. Does this invalidate Pink's thesis? It certainly dates the way he has chosen to express it, but as it happens, many of his arguments can still be taken as valid so long as, each time you read 'right-/left-brained', you take it to mean 'people who are strong in attributes traditionally associated with either the right- or the left-brain'.

Pink points out that the professions our parents traditionally wanted us to go into – law, medicine, accountancy, engineering – are all based on 'left-brain' attributes. Now the situation has changed: we are entering the 'conceptual age' and companies of the future will need designers, creators, trainers, and holistic visionaries – all jobs that suit 'right-brain' thinkers.

The theory of right-versus-left brain is described in some detail, and this is pretty interesting, so long as you remember that pinch of salt we started with. This leads into the core of the first part of the book: a discussion of the three major reasons why left-brain attributes are continuing to become more important – 'abundance, Asia and automation'.

Abundance is more or less self-explanatory. We live in an age when we have a wider than ever choice of any goods or services you can think of. 'Abundance has brought beautiful things to our lives, but that bevy of material goods has not necessarily made us much happier.' As a result, Pink argues, people are increasingly searching for meaning and purpose, best served by left-brain attributes.

The issue with Asia is that there are generations of young students from areas that include China and India who are now as well-qualified and educated as the young of the west. We live at a time when huge amounts of the work that is currently carried out in the west can be outsourced to those countries.

Automation is introduced with examples of humans being out-skilled by machines, from John Henry's fatal race against a steam engine to Gary Kasparov's failure to defeat the supercomputer Deep Blue. The moral is obvious: any job that can be reduced to lines of code and mechanical systems can and probably will eventually become automated. Only jobs that can't be automated are future-proofed.

The second part of the book explores the six essential areas of expertise that will be needed in the future: design, story, symphony, empathy, play and meaning. Pink devotes a chapter to each of them.

Design is crucial because of the economic value of making something beautiful or emotionally convincing. Story under-pins all attempts at persuasion and communication: 'I say, "Get me some poets as managers." Poets are our original systems thinkers.' Symphony refers to the importance of seeing the big picture, and synthesising ideas and elements. Empathy allows an individual to truly understand the choices and preferences of others. Play (as opposed to sobriety) is a crucial element of making the workplace (and thus the products) fun and arresting. And meaning is the solution to the emptiness of a world of abundance and consumption. Each of these chapters ends with a 'portfolio' of exercises and tools, along with further reading.

Pink is an engaging writer and there are plenty of fun facts to keep the reader moving forward. For instance, he analyses the physiological difference between a sincere smile and an insincere one, and describes a visit to a 'laughter club' in Bombay. The left- and right-brain theory he bases the book on may be flawed, but the challenges he describes are real ones that we are still facing today.

THE SPEED READ

Different individuals have right-brain and left-brain attributes. Where logic, verbal ability and linear reasoning (left-brain attributes) were once the key to business success, now we face the combination of automation, an abundance of goods and services and the ability of the west to outsource elsewhere in the world. The attributes that will preserve western jobs will be right-brain attributes such as creativity, holistic thinking and vision.

Whatever You Think, Think the Opposite

Paul Arden, 2006

This is a nicely designed little book that hammers home the idea that contrarian thinking can conquer conventional ideas.

Arden was an executive creative director at the legendary advertising agency Saatchi & Saatchi and it shows, both in the flashy design – heavy on images – and the flashy typography – emphasising elements in the brief text.

Arden's work retells stories of people who think outside the box. For instance, he discusses the way that Dick Fosbury reinvented the high jump at the 1968 Olympic Games. There had been only three techniques known by high jumpers: scissors, western roll and straddle jump. But Fosbury came up with his own unique method, turning to jump backwards over the bar at the last moment. It became known as the Fosbury Flop and it won him the gold medal, as well as allowing him to set a new world record.

We also hear of Allen Lane, who set up Penguin in 1934: no one believed that selling cheap paperbacks was a feasible business plan, and most booksellers weren't interested. It was only the fact that Woolworths committed to taking the books that allowed him to try, but his publishing list ended up being one of the most successful business ventures ever. Lane's story confirms Arden's assertion, 'The world is what you think of it, so think of it differently and your life will change.'

There are some nice touches that illustrate different ways that you might challenge paradigms. For instance, Arden discusses, using a visual approach, the many different ways you might photograph a flower. These include capturing wilted flowers, empty vases, and extreme close-ups that transform the way we think about flowers.

The book is genuinely inspiring and nicely put together. The problem is that there is a limit to the number of different ways the thesis can be stated even in a small picture book. Once you've said, 'If you always make the right decision, the safe decision, the one most people make, you will be the same as everyone else,' you need to find different ways to say the same thing. There is a tendency in the book to drift into wild overstatements such as, 'It's wrong to be right; it's right to be wrong.'

There is also an unsurprising tendency for Arden to come across as egotistical: Saatchi & Saatchi were notoriously hubristic and self-promoting in his era, and it isn't surprising that the author of *It's Not How Good You Are, It's How Good You Want To Be* might be a touch self-regarding. When he says, 'Great people have great egos; maybe that's what makes them great,' it's hard not to wince and to assume he is partly talking about himself.

If you can tolerate this apparent narcissism, Arden's book is a useful pep talk and resource for reminding yourself that 'experience is built from solutions to old situations and problems. The old situations are probably different from the present ones, so that old solutions will have to be bent to fit new problems (and possibly fit badly).' If you want to be prodded into taking more risks and thinking in different ways, you could do worse than leave a copy of this book on the coffee table or in the bathroom.

---THE SPEED READ---

Don't think the way other people do or the way you did in the past. Do something amazing and original that everyone will tell you is wrong. But you'll be right. Because it's always right to be wrong. Seriously, take more risks, forget about the past, and believe anything is possible and you might just change the world.

What I Talk About
When I Talk About Running

Haruki Murakami, 2007

As a novelist Haruki Murakami is like Marmite: loved or hated. Novels such as *A Wild Sheep Chase*, *The Wind-Up Bird Chronicle* and *Hard-Boiled Wonderland and the End of the World* create strange, overlapping versions of the real world and alternative realities. Enigmatic characters wander through in seemingly random plots. The effect is somewhat like listening to the improvised jazz which Murakami loves, or less scary versions of David Lynch films.

Murakami also writes on a variety of non-fiction subjects, and runs marathons and triathlons in his spare time. As with everything he does, he takes a somewhat obsessive approach to his running. A lot of the interest in this book (which was first published in the English language in 2008) comes from the ways in which the author links running to writing and the creative process. He argues that both running and writing rely on three main qualities – in descending order of importance – talent, focus and endurance. He is somewhat elusive on whether or not the running directly influences his writing. At one point he notes, 'Most of what I know about writing I've learned through running every day.' But elsewhere, 'Occasionally, hardly ever really, I get an idea to use in a novel. But really as I run, I don't think much of anything worth remembering.'

It seems that it is the discipline and persistence that Murakami applies to his running and the sheer level of ritual it involves which is what connects with writing and allows him to return to his work each day. There are also some interesting insights into the way the mind works: as he runs he ponders the fleeting nature of existence, and also notes the way that thoughts are popping in and out of his mind as he runs, like clouds in the sky.

Murakami's writing style can be mundane and plain at times and whether or not you find this charming or irritating will be down to your literary taste. A lot of what he talks about here is fairly ordinary: the fact that he runs to avoid things and to avoid getting fat; the way that people who ask what he thinks about when he is running are always people who have no experience of running themselves, so no experience of the way the mind wanders around; the type of rock music he runs to, which he chooses for the right rhythm.

Alongside this there are interesting meditations on ageing (as he had turned fifty when he wrote the book) and the motivation of the runner. He is someone who naturally likes being on his own, and who isn't competitive, except with himself, which is what spurs him on when he is on a run. Other aspects of the book will, obviously, have a special appeal for readers who are runners rather than writers or philosophers. The description of Murakami running from Athens to Marathon with only a photographer for company is particularly striking.

On the whole this isn't a revelatory read, but it is an oddly reassuring, human book that says a lot about how this particular writer's mind works. He puts his thoughts and motivations in a way that challenges the reader to think about their own.

THE SPEED READ

The brilliant Japanese writer discusses his enthusiasm for running and draws out comparisons with the creative process, while giving an insight into what drives talented people on to show the focus and persistence required to complete either a marathon or a long novel.

Predictably Irrational:
The Hidden Forces That Shape
Our Decisions

Dan Ariely, 2008

One of the problems with traditional economics is that it assumes that people make rational decisions. As scientific writers including Daniel Kahneman have demonstrated, our choices are often quite irrational.

Dan Ariely is an MIT professor of behavioural economics. His book is a detailed look at the irrational forces that drive our decision-making, including a plethora of fascinating anecdotes and reports into experiments that demonstrate our systematic biases. The point of the title is that we often behave in irrational ways, but our biases are sufficiently well-understood that it is often possible to predict exactly how irrationally we will behave.

One of the most compelling revelations in the book is the degree to which our behaviour is affected as soon as we perceive it in economic terms. Ariely writes about a situation in which lawyers were asked if they would offer advice to retired people at thirty dollars an hour. There were no takers. However, when lawyers were asked to do the same thing for free, most of them volunteered. Ariely explains this peculiar phenomenon thus: 'When money was mentioned, the lawyers used market norms and found the offer lacking, relative to their market salary. When no money was mentioned they used social norms and were willing to volunteer their time.'

Similarly, when a group of students were paid to help with a research project, the subgroup who were paid more worked harder than the subgroup given a lower rate. However, another group who volunteered with no reward worked harder than all of them: again, they were applying social norms, whereas market norms were affecting the perceptions of both paid groups.

Another fascinating observation is the power of something being 'free'. When a test group was offered Hershey's chocolate for one cent and the far superior Lindt chocolate for fifteen cents, most chose the better-quality chocolate. However, once the price of both items was reduced by a cent, most people chose the free chocolate. This explains the attraction of 'buy one get one free' offers over two goods at half-price and the proportion of people who buy two books from Amazon, to get free delivery, in spite of not really wanting the second book.

There are a few psychological phenomena given extensive treatment in the book. 'Anchoring' is a name for the way that precedence can influence our view of what constitutes a reasonable price. A forty-dollar bottle of wine will look overpriced next to ten-dollar bottles on a menu, but we may see it as a bargain if it is next to wine costing two hundred dollars. Ariely points out that Starbucks managed to increase our anchoring price for a cup of coffee. By creating a very different environment in which to sell their coffee they managed to avoid the precedence effect of other coffee shops lowering their own prices. Once people became used to the concept of paying the higher price, the cost went up across the board because of this anchor.

As with other cognitive biases, anchoring can be hard to escape, but one way is to think of the price out of the current context. If you're offered a more expensive bottle of wine, think about what that money would buy you in a different situation and then compare the satisfaction you will get from the two alternatives.

Another bias covered in detail in the book is the effect of our expectations on our perceptions. If we take an aspirin costing a penny, it won't have as big an effect on our headache as one that costs fifty pence – at least that is how we will perceive and report it. Students offered a free sample of beer were more likely to report it tasted bad if they were told beforehand it had had a drop of vinegar added to it than if they drank the exact same beer without being told this.

Finally, Ariely takes a cold, hard look at how our morality functions in real-life situations: 'When given the opportunity, many honest people will cheat,' writes Ariely. He sets out to explore the factors that affect our behaviour when offered the opportunity to do so. Insights here include the fact that we are less willing to steal cash from a fridge than a case of beer of the same value, but that our reluctance is more muted if instead of cash they left tokens in the fridge (which can later be exchanged for cash).

One experiment not for the squeamish involves an explor-ation of how sexual arousal affects our morality. Participants are asked a list of questions in two scenarios about their sexual limits and parameters. First, they are asked the questions in a normal setting. Then they are asked the exact same questions while sexually aroused. (Let's just say that they are in front of a computer which has had its screen and keyboard clingfilmed by the fastidious experimenters.) The result is that the participants report their moral boundaries to be much broader when they are in a state of arousal than when they are not. There is a slightly jarring note of sexism in the way this experiment is set up and reported, but the results are nonetheless interesting.

The final cognitive bias to feature prominently in the book is the endowment effect, in which we value things we already possess more highly than things we don't. Ariely uses many personal stories throughout and talks of the time he and his wife tried to sell their house and were forced to face the sobering reality that other people simply didn't rate it as highly as they did.

There are briefer accounts of the field of behavioural economics – Ariely is prone to giving six examples where one or two would have made the point. Other than that minor flaw, this is an excellent book that reveals those hidden forces that affect our choices, and gives good advice on how to counteract those biases.

THE SPEED READ

We imagine that we are rational beings, but our economic choices, in particular, show that we aren't. Simply participating in an economic situation is enough to affect our behaviour, as we choose market norms over social norms, and almost every time we value something we are being affected by factors that we haven't recognised.

Your Brain at Work

David Rock, 2009

When you sit down to work, whether it is at home or in the office, do you find it hard to maintain focus? Do you end up surfing the net, checking your phone obsessively, wandering off to chat, or watching daytime TV in your pyjamas? (Honestly, this isn't an autobiographical book . . .) If so, this is a useful read. *Your Brain at Work* focuses on how to deal with the flood of interruptions, high stress levels, poor self-control and excessive expectations.

David Rock is a neuroscientist and an expert on high performance. His background leads to some pretty technical advice along the lines of, 'May your cortisol levels stay low, your dopamine levels high, your oxytocin run thick and rich, your serotonin build to a lovely plateau, and your ability to watch your brain at work keep you fascinated until your last breath.' Fortunately, he does find ways to translate this level of scientific understanding into everyday language and to give practical advice about what is actually going on in your head when you are easily distracted.

Rock uses two fictional characters, Paul – who works both at home and on contract in offices – and Emily, who is VP of marketing for a corporation. They have complicated lives and he recounts the various challenges they face throughout their day. He discusses their emotional challenges, the ways they try to regain focus and how they need to deal with other people, giving constructive feedback when needed, among all their other tasks. He also talks about the importance of breaks in refreshing their attention.

Rock restresses the commonly held idea that multi-tasking is a myth: your ability to think is limited and you only lower your performance if you try to do many things at once. You can end up acting as though your IQ is ten points lower than it is. He explains that the prefrontal cortex is the engine of our mind and there is a limited time when it can perform at its best. If we want to access that sweet zone of productivity, we need to manage our mental energy carefully.

Just as you need to rest, sleep and recover, you also need to ration the time you spend problem-solving. A group of subjects back in the nineteenth century was asked to push a dynamometer, which measures the force being exerted on it, while solving a mental problem. The harder they had to think about the problem, the less force they exerted. Their performance on the physical task ended up as much as 50 per cent lower.

Rock's main suggestions for combating this reduction in brain-power is first to be ruthless in prioritising your tasks, in the morning, when your brain is at its most adept, and to learn to convert your most important tasks into habits, allowing you to perform them on autopilot, thus using less mental energy. He also suggests competing against yourself, or at least against the self you were yesterday.

Status, and how we feel about ourselves, can have a physical effect, whether it is being congratulated on a great performance, buying a fancy pair of shoes, or achieving a goal. The elevated sense we have of ourselves will lead to higher dopamine and serotonin levels, making us happier, and lower our levels of the stress hormone cortisol. The importance of this isn't restricted to how it makes us feel. The same hormone levels will stimulate our neurons to make faster connections, boosting performance.

The key thing is that you can achieve this effect simply by beating your previous performance. If today's self beats yesterday's then today's self gets a little ego boost that helps trigger better happiness and quicker thinking.

Rock's advice on how to give constructive feedback is especially interesting. He points out that people only truly understand a solution and identify it if they come up with it themselves. This may be a self-centred (albeit natural) trait, but you can make use of it. Giving people advice only works a small percentage of the time, but if you can help steer someone in the right direction so they find the solution themselves, you will have a greater effect. You also reduce another person's stress levels if you act as their thinking coach rather than as their instructor.

Some readers find the stories about Paul and Emily a bit folksy and simplistic, but Rock's broad knowledge of how the brain works and his ability to explain it in clear terms is a valuable thing. Next time you find yourself struggling to concentrate on a task, refresh your mental energy by taking a short break and ordering a copy of this book.

THE SPEED READ

In order to get better at focusing on work and to overcome the challenges of distraction, you need to understand what is actually going on in your brain. Conserve and ration your mental energy by ruthlessly prioritising and turning key tasks into habits. Give yourself an easy win by competing with your yesterday self. And learn how you can help people help themselves while letting them think it was all their own idea.

Making Ideas Happen:
Overcoming the Obstacles Between Vision and Reality

Scott Belsky, 2010

This book claims to be based on interviews with hundreds of the most creative people and teams in the world, the sort of suggestion that can provoke scepticism in the more cynical kind of reviewer. Author Scott Belsky, however, is founder of the online creative community Behance and is well-placed to share his views of what makes the difference between being creative and *productively* creative.

He comments in the introduction that creative professionals are 'responsible for the designs, entertainment, literature and new businesses that bring meaning to our lives'. At the same time, they are some of the most disorganised people on the planet. His aim is to offer advice on how to overcome that innate disorganisation and increase output. To this end he also focuses on serial creatives: people who have a history of having ideas and making them happen. His observation of serial creatives has led him to explore some of their most common, prominent features: they share a high level of organisation; they interact with their peers and use the power of community as a source of support and momentum; and they have good strategies for building and leading creative teams.

There are three main sections in the book. The first focuses on organisation and execution: Belsky writes about the 'action method' that he developed to help creative people organise their projects as a way to keep them focused on next steps and the processes they need to complete to finish a project. (Jargon alert: if you are allergic to business-speak you may find this book painful at times.)

The action method relies heavily on keeping notecards in a system in which each project has its own card. You break the project down into 'action steps', 'references' and 'backburner' and then make a note of the dates on which you have taken action. He also suggests using an 'energy line', a scale going from 'idle' to 'extreme' (via 'low', 'medium' and 'high'). You might, for instance, have this line on a whiteboard with a sticky note for each project, and move the projects up and down as required.

He is also firm in his advice that you need to know when to give up on dead-end projects. This is something many creative people find hard to do, as they are afflicted by a version of the 'sunk costs' fallacy, unwilling to accept that effort might have to be written off. As Belsky writes, 'You must be willing to kill ideas liberally – for the sake of fully pursuing others.'

The second section of the book focuses on community. With his background in Behance, Belsky is obviously an evangelist for publicising your ideas to gain support, feedback and constructive criticism. He emphasises the important role that can be played in the creative process by partnerships and mentors and how much more effective teamwork can be than being a lone wolf. Later in the book he talks of 'visionary's narcissism', a state in which creatives become so fixated on their idea and so convinced it is unique that they don't even research the competition. In contrast he suggests that 'watching your competition . . . can be a great way to push your idea into action . . . seek out competition and be grateful for it.' Going public about work also creates accountability and commits you to the process: 'only after publicly committing themselves did [leaders of new companies] experience full support from their communities.'

The final section of the book focuses on leadership and self-leadership. Belsky emphasises the latter because, as everyone knows, we can be our own worst enemies. But he also talks about the process of building a creative team, finding people with complementary skills and motivating and rewarding them. He focuses on the way that a good team works effectively as an 'immune system' that will reject bad ideas. There is also an interesting passage on seeking out the hot spots in a company. A study of one Fortune 500 company polled the employees to find out who they turned to for advice when there was a problem they needed to deal with, whether it be creative advice, a computer glitch or a finance problem. This was mapped onto a diagram with each employee represented as a node in a network. The fascinating result was that the resulting map bore little resemblance to the job titles, level of seniority or even area of expertise of the employees. The conclusion was that companies are probably far more reliant on their hot-spot employees than they realise. By searching out the people in a company who have social power, and giving them responsibility and involving them in projects, Belsky suggests you can significantly improve the output of your creative team.

Different people take different approaches to the problems discussed in this book: many readers react positively to David Allen's *Getting Things Done*, while others find it to be a weirdly neurotic time management system that will take so much of your energy it will prevent you from actually getting anything done at all. Belsky's book is more focused on the workplace and the creative process in general and – if you can stomach a high level of business jargon – it is a valuable book that contains many moments of insight.

┌─ **THE SPEED READ** ─────────────────────────┐

Creative people are chaotic and also some of the most important individuals in any company. To maximise your output as a creative you need to adopt the 'action system', breaking down jobs into concrete steps and making regular notes of progress. You also need to access the power of community in developing your ideas, be willing to drop bad ideas, and learn the power of self-leadership as a means to improving your leadership skills.

└──┘

The Shallows

Nicholas Carr, 2010

An older colleague recently commented that young people seem to have lost the ability to navigate their way around their own cities. The theory was that they have got used to relying on maps apps on smartphones and satnavs, with several consequences. They don't really look up and actively engage in the route they are taking, they just follow like lab rats. They don't spend time studying a map and memorising or internalising crucial parts. And they simply don't look around and take in the city as much as they might because, even as they are walking, they are checking their phones and texting their friends.

This observation fitted in nicely with the thesis of Nicholas Carr's 2008 article in the *Atlantic*: 'Is Google Making Us Stupid?' That was the basis from which he expanded his argument to book length. The essence is that the digital age is having a profound effect on many elements of the way our minds work. 'We become, neurologically, what we think.'

Carr bases this observation in a fascinating exploration of changes in human technology and the way they have affected our thinking. From alphabets to counting systems, maps to the printing press and from the clock to the mass media: every one of these inventions has allowed us and required us to think about the world. Carr refers back to the work of Marshall McLuhan who argued, 'The medium is the message'. The point is that, whether you take in information through books, newspapers, TV or the computer, the medium itself is hugely important in shaping and packaging the information.

Carr discusses the recent work of scientists such as Michael Merzenich and Eric Kandel in introducing the concept of 'neuroplasticity'. This is a relatively recent theory: our brains continue to adapt and change (in a 'plastic' way) throughout our lives. Our experiences and sensations are part of the way in which this happens. When we use technology of any sort – ancient or modern – to use, store or share information, it literally reroutes our neural pathways.

What does this mean in terms of the digital world (that Carr contrasts most strongly with the era of books). Well, for a start, it means we are getting used to simple methods of finding, skimming and extracting information: we no longer have a need to memorise information, and we access it in quite a scattergun way. Carr suggests this is affecting our ability to concentrate on and contemplate information. 'Whether I'm online or not, my mind now expects to take in information the way the net distributes it: in a swiftly moving stream of particles. Once I was a scuba diver in the sea of words. Now I zip along the surface like a guy on a jet ski.' He sets this up in contrast to the literary experience: 'In the quiet spaces opened up by the prolonged, undistracted reading of a book, people made their own associations, drew their own inferences and analogies, fostered their own ideas. They thought deeply as they read deeply.'

While Carr is not entirely negative about the consequences of our computer use, he can be quite doomy. He talks of how the internet is essentially a disruption engine. 'We want to be interrupted, because each interruption brings us a valuable piece of information. To turn off these alerts is to risk feeling out of touch, or even socially isolated.' He also suggests that our addiction to the flood of information available online is affecting our ability to read and enjoy books, and is thus limiting our mental abilities. (There is some experimental evidence that casts doubt on this theory, showing that heavy internet users are still able to process the information presented in books as well as non-users, but at the same time the falling rates of literacy among the young provide at least circumstantial evidence to support the idea.)

There is an irony here. The medium is indeed the message, and the writing of a book (as opposed to an article) requires a word count large enough to justify the cover price imposed by the publisher. In all too many cases books of this sort end up being bloated, as they stretch their main idea out to fill the space, and there is a degree to which that seems to have happened here. For all the amusing vignettes of Nietzsche's frustrations with a typewriter and Freud's dissection of maritime creatures this is still, at heart, an op-ed padded out to make a book. The level of repetition can lead one to wonder how scientific its basis truly is. But for all that, this is certainly an interesting exploration of the way that our devices and technologies have moulded our brains and are continuing to do so via the internet.

┌─ **THE SPEED READ** ─────────────────────────────┐

Our brains are moulded by the devices we use to access information: books encourage reflection and deep learning, whereas the internet allows easy skimming of information, which we therefore never fully contemplate, concentrate on or store. As a result we are getting stupider and less able to focus on book-length expositions. (And never ask a twenty-something to navigate without their phone.)

└──┘

The Art Of Thinking Clearly

Rolf Dobelli, 2011

Rolf Dobelli is a Swiss author and when it came to thinking about how to protect the modicum of wealth he had built up, he decided he needed to ponder how to avoid making stupid, avoidable mistakes. This train of thought led him on to writing this book (first published in English in 2013), which is essentially a guide to cognitive biases and how to avoid them.

You might not feel the need to read this if you have read books like *Predictably Irrational* (see p.114), *Thinking, Fast and Slow* (see p.144), or the oeuvre of Nassim Nicholas Taleb. But this is an admirably concise attempt to address many of the same topics. It is a relatively short, elegantly written book, containing a hundred chapters, mostly about two to three pages long, each covering a specific cognitive bias and suggesting remedies.

Dobelli starts with a brief explanation of cognitive error: a systematic failure to apply optimal rational thought or behaviour in a given situation. His suggestion is that we learn to spot these errors and our tendency to repeat them, and we can then hugely increase our levels of success in life. We need only cut out the irrationality and that doesn't even require new skills, only that we treat our own rational ability with more scepticism. The great thing is that many ways of achieving this do make real sense, and are not that difficult to apply.

For instance, he briefly describes survivorship bias, in which our understanding of the likelihood of success is biased by the fact that the people we are most aware of in any given industry or role are those who have had the luck or skill to thrive over a long period. He suggests we continually remind ourselves of past failures, revisiting the 'graves' of decisions, careers and investments that went catastrophically wrong.

When it comes to our tendency to look for patterns and to see them where they don't exist, his advice is to always be sceptical if we believe we are seeing a pattern, and to start from the assumption we are probably wrong if we do. If you are really finding it hard to shake the idea you've spotted a pattern everyone else has missed, then it may be a good idea to consult an expert in statistics.

One of the most important pieces of irrationality, in his view, is confirmation bias. This is extremely common in business when, for instance, any sign of a strategy being successful is welcomed with enthusiasm, whereas evidence it is failing is downgraded or ignored. He suggests writing down your initial beliefs about all sorts of aspects of your life, from your health to your relationships, and from your business life to your personal life. Then you should actively look for information that contradicts these beliefs. Only by doing that can you establish a truly independent, unbiased view.

Another common misconception is what Nicholas Nassim Taleb has called the 'swimmer's body illusion'. This comes when we mix up selection with other factors, just as when we believe that successful swimmers develop their bodies through the exercise, rather than understanding that a certain physique is a prerequisite for becoming a strong swimmer in the first place. This can apply in many different situations: for instance, consider the question of whether or not Cambridge is a good university with excellent teaching. It is hard to truly establish this. We know that its reputation means that it is guaranteed to attract some of the most intelligent, highly motivated applicants, and they would be most likely to do well in life regardless of whether they are taught well or not.

Dobelli also follows up Taleb's well-known 'black swan' analogy (assuming that all swans are white because you don't live in a country where the black variant can be found). In this case his advice is rather conservative, suggesting that you avoid any venture that could be affected by a black swan event and keep your investments and business decisions on the safe, cautious side.

Dobelli uses the famous Solomon Asch experiment as an example to remind us of the power of peer pressure. In this test, participants are shown a line and asked to compare its length to three lines drawn next to it. One is shorter, one is the same length, and the other is longer. If they are asked on their own to choose the correct line (the one that is the same length) they respond correctly. However, if five other people give their judgment first – all actors, primed to assist the experimenter – and choose the shorter line, a significant number of participants will be affected by the pressure. They succumb to choosing the 'wrong' option.

There are too many cognitive errors discussed here to list, including the sunk costs fallacy, hindsight bias, availability bias and many more. Indeed, there are so many different names for biases that one occasionally wonders if Dobelli is being a bit pedantic in describing them all as separate. Some readers also find the author a bit self-congratulatory and pompous at times. But that is a minor cavil. On the whole, this is a terrific book, and a great potted survey of the cognitive biases we all suffer from, as well as a guide to combating them.

THE SPEED READ

Cognitive biases are those inbuilt tendencies that lead us to think irrationally, leading to errors like making unwise eBay purchases, bad investments, and much more besides. Here is a short guide to help you recognise and overcome all kinds of cognitive bias.

Little Bets:
How Breakthrough Ideas Emerge From Small Discoveries

Peter Sims, 2011

Peter Sims, a former venture capitalist, starts this book by talking about the way the comedian Chris Rock puts together a touring performance.

First, he turns up unannounced to a comedy club near his home. Rather than going into his usual full-on preacher effect, he takes a quieter, more conversational approach, trying out reams of new jokes, and occasionally breaking off to make notes on the yellow notepad he has brought with him, from which he is often reading. Many of the jokes fall painfully flat, but a few here and there are home runs which bring the audience down. Sims' point is that it can take a lot of these kinds of small steps to work your way up to a big result. By taking little bets, small risks that won't cost Rock too much if he fails, he can test out his target audience's response to particular jokes and even to particular ways of telling jokes.

A similar story can be told about the growth of animation production company Pixar. It was a computer hardware company at the time that Steve Jobs bought it. He didn't foresee

Pixar becoming a success in the field of animation, but he was willing to allow John Lasseter and a small team to start out making short films, as brief as a minute long, to see where it led them. Which was, of course, to some of the great animated masterpieces of the digital age, including the *Toy Story* movies, *Finding Nemo* and *Wall-E*.

Not everyone needs to be a prodigy like Mozart, who could simply make a single leap to a brilliant idea without experimentation or failure along the way. Yet people expect that they have to do this at the start of any business venture. Sims writes, 'The seed of this book was planted while I was attending Stanford business school. One of the most common things I would hear people say was that they would do something new – take an unconventional career path or start a company – but that they needed a great idea first.' Having worked in venture capital, Sims knew that many entrepreneurs who succeed discover their brilliant ideas on the journey rather than starting with them.

Rather than focusing on the possible end gains, start out by focusing on what you can afford to lose right now. Every time you have an attempt at an idea, no matter how small, start road-testing it. Sims advises talking to as many people as possible: all entrepreneurs need curiosity, to find out people's opinions. You can learn a lot from even a small amount of material, before improvising, testing, experimenting and reiterating the process. The story of Thomas Edison's ten thousand failed attempts at making a lightbulb is a bit of a cliché but it makes perfect sense in relation to this book.

Start out with a growth mindset (see p.152 *Mindset*, Dweck) with which you see failures or near misses as an opportunity to learn. This doesn't mean you can't later aim for a level of perfectionism, simply that you don't let perfectionism paralyse the creative process right at the beginning.

This also has an impact on the way that managers seeking to foster innovation should treat their workforce. A playful atmosphere in which experimentation is encouraged, and seen as a good thing, and in which people aren't chastised but praised when they fail in their little bets will help your company to find creative solutions. The innovative manager needs to fully immerse themselves in the process, questioning, observing, and challenging at each stage.

Sims also points out that, once you embrace the idea of little bets, you can incorporate them into every stage of the process. Even when a product is close to being ready for market, you can use little bets to test out packaging, marketing and sales ideas. And you can continue to use them to test out possible ways your product could evolve and change after reaching market.

Given his background, it is not surprising that Sims mostly focuses on the business applications of his idea, but it is something that could be adapted for all kinds of situations. An artist trying out a new style could experiment with sketches, or a parent could see what happens if they try a different method of persuasion on a recalcitrant child.

THE SPEED READ

Rather than expecting brilliant ideas to come out of nowhere, you need to start with little bets. These are experiments or trials of imperfect ideas, undertaken in order to invite feedback and to take the first steps towards success. If you continue to apply these throughout the process, your brilliant idea is more likely to emerge on the journey.

Moonwalking with Einstein:
The Art and Science of Remembering Everything

Joshua Foer, 2011

Before writing this book, journalist Joshua Foer had a crash course in the art of memorising that led to him becoming US memory champion in 2006.

Foer was guided in his quest by K. Anders Ericsson, a Swedish psychologist who specialised in the field of expertise and performance. As Ericsson's book *Peak* will be covered later (see p.240), we will mainly look here at the parts of Foer's book that focus more on the memory techniques than on the parts that involve pushing yourself beyond the 'plateau' to become an expert. It is worth knowing that Foer generously credits Ericsson with many of the insights he is sharing here.

The basics of becoming a mnemonist, someone with the ability to memorise large amounts of data, are well known, and go back to the ancients. In the fifth to sixth century BCE Simonides of Ceos reportedly used a method called loci, also known as the 'memory palace' technique. It consists of associating each piece of data you want to remember with a memorable image and 'placing' that image in a specific location in an imaginary version of a well-known building. In order to retrieve the memories, you travel through the building in a predetermined direction and 'find' the memories, one by one. The title of the book comes from one of these bizarre juxtapositions of images and Foer points out that, as our brains seem programmed to have a fascination with sex, having a dirty mind can help. Lewd or bizarre images associated with a piece of information may help us to remember it.

The method is also recorded in the *Rhetorica ad Herennium*, the oldest surviving Latin book on rhetoric. Ancient mnemonists such as Cicero made a distinction between *memoria verborum* (memory for words) and *memoria rerum* (memory for things). The latter is easier to establish, and association methods tend to help. Foer explains, 'As bad as we are at remembering names and phone numbers and word-for-word instructions from our colleagues, we have really exceptional visual and spatial memories.'

We don't tend to memorise isolated facts, but rather store them in context. Even details such as where we first learned a fact can help us to retrieve it. It also helps if we can find ways to break information into chunks. It's easy to remember long phone numbers, for instance, if we can 'chunk' them. In this way, 0738932719 will be easier to remember if we chunk it as 07 38 93 27 19. Advanced memory systems include ways of converting numbers and words into visual cues.

Familiar as this is, Foer's sheer enthusiasm for his task makes this an entertaining read: the book has been optioned for a movie and one can imagine the story of a breathless geek becoming a national champion with the support of a (probably) eccentric psychologist being entertaining.

The other interesting aspect of the book is the more philosophical side of Foer's thoughts on memory. He sees memory as the key to a good life. 'The more we pack our lives with memories, the slower time seems to move. Monotony breeds fast time.' He advises the reader to pack their lives with experiences, travelling to different locations, and notes that your life will not be memorable if you spend too much of it sitting in the same cubicle at work. 'Creating new memories stretches out psychological time, and lengthens our perception of our lives.'

He also notes how much the 'externalisation of memory' through social media, photography, video and so on has altered our relationship with memory. It can seem less important to commit things to memory in a world where everything can be recorded or googled. 'So why bother investing in one's memory in an age of externalised memories?' And he describes the experience of someone who lost their memory, and literally couldn't place themselves in the world. 'How we perceive the world and how we act in it are products of how and what we remember. We're all just a bundle of habits shaped by our memories.'

In the end, as Foer says, learning to remember, and having experiences that you want to remember, are essential elements of a good life. If the book were merely an amusing story of a geek learning a new skill, it would still have some virtue, but the charm of the author and this philosophical strain add extra ingredients, making this a most satisfactory concoction.

---THE SPEED READ---

I want to learn how to be a mnemonist, so I approached a world expert in performance. He helped me train so effectively that I became a memory champion. But I also learned how important memory is in defining us as human beings, and how important it is not to let go of that in an age of the externalisation of memory.

The Psychopath Test

Jon Ronson, 2011

In his twenties, Jon Ronson was the keyboard player for Frank Sidebottom, the alter ego of Manchester singer Chris Sievey, who developed the character while wearing a huge, papier-mâché head. When his head was in place, Sievey he would answer only to the name Sidebottom.

Since then, Ronson has written books about extremists, and theorists of conspiracies and military telekinesis. As a result, he has experience of hanging around with some strange people, and has a fairly clear understanding of what makes them tick.

It is not so much of a surprise to find that the research for this book included interviews with characters as varied as the leader of a Haitian death squad, a notoriously odd CEO, a patient in an asylum for the criminally insane who argued he was in fact sane, and a Scientologist working in the anti-psychiatry wing of the movement. The inspiration to write the work had come when, while researching an obscure academic hoax, Ronson came across the Hare psychopathy checklist. Devised by Bob Hare, a Canadian psychologist, this is a list of questions to identify traits that are generally displayed by psychopaths: glibness, superficial charm, a grandiose sense of self-worth, pathological lying, cunning and manipulative tendencies, a lack of remorse, shallow affect (an inability to feel deep emotion), callousness, lack of empathy, and a failure to accept responsibility.

Ronson looks into the idea, expressed in the book and film *The Corporation* and elsewhere, that many high-flyers in business are actually psychopaths, whose condition is responsible for their success. Indeed, *The Corporation* takes this idea further, arguing that corporations are, by design and definition, psychopathic in their very essence.

He is also interested in how broad and loose definitions of mental conditions are, and interviews Anthony Maden, a psychiatrist at the dangerous and severe personality disorder (DSPD) unit at Broadmoor. Maden points out that someone coldly planning a project or someone highly impulsive could end up scoring high on the Hare test, and uses this to express doubt about the value of such checklists.

This doubt takes on a much broader significance when Ronson examines the *DSM* (*Diagnostic and Statistical Manual of Psychiatric Disorders*), a catalogue of conditions published by the American Psychiatric Association. It defines a huge range of conditions in fairly narrow terms, with some pretty odd inclusions, such as intermittent explosive disorder (which could be called having a short fuse and a violent temper), and sluggish cognitive tempo disorder (which could just describe someone who is lazy or unmotivated). Ronson's concern is that behaviours exhibited by a huge number of ordinary, 'sane' people could be classified as symptoms of a condition, according to the *DSM*.

'[When] I asked Robert Spitzer [editor of the catalogue] about the possibility that he'd inadvertently created a world in which ordinary behaviours were being labelled mental disorders, he fell silent,' writes Ronson. 'I waited for him to answer. But the silence lasted three minutes. Finally, he said, "I don't know."' This is a pretty astounding response, especially given the huge rise in the prescription of drugs for adults and children with conditions such as bipolar disorder and ADHD (attention deficit hyperactivity disorder).

Ronson also reflects on how flexible the idea of madness can be, noting that in the 1850s the American doctor Samuel Cartwright 'discovered' a condition known as drapetomania. It was only found in slaves, the symptom was 'the desire to run away from slavery' and the cure was whipping.

Of course, a book of this sort can lead one to speculate on one's own sanity as well. Ronson helpfully reports psychologist Martha Stout telling him, 'If you're beginning to feel worried that you may be a psychopath, if you recognise some of those traits in yourself, if you're feeling a creeping anxiety about it, that means you are not.'

On a broader note, Ronson laments how often people have become defined by their 'maddest' traits and how often life in the media involves playing up those traits to become a caricature of a real person. In addition, he wonders why so many health professionals who are responsible for diagnosing mental illness seem to be quite strange themselves.

This all makes for a lively and very funny meditation on the nature of madness and sanity. The main problem is that Ronson tends to be glib and often skates over issues that deserve much deeper thought, in order to reach the next wacky interview, and another opportunity to insert himself into the narrative, a key feature of his writing. But the book is great fun and if it provokes more questions than it is able to answer, Ronson is at least brave enough to ask these difficult questions in the first place.

THE SPEED READ

Think you're crazy? Then you're probably not . . . The standard test for psychopathy is sufficiently flexible to characterise many different personality types as psychopaths, and there are similar problems with the definition of many other mental conditions. Jon Ronson tours the world, speaking to an eccentric and fascinating range of characters, to explore the idea of madness and sanity and to ask some of the questions that the mental health industry would prefer to ignore.

Thinking, Fast and Slow

Daniel Kahneman, 2011

Daniel Kahneman is a Nobel Prize winner who has carried out decades of brilliant research into psychology and behavioural economics, mostly in collaboration with his colleague Amos Tversky, in whose memory this book is dedicated. The largest part of Kahneman's psychological work focuses on cognitive biases and false beliefs and conclusions that humans are nevertheless prone to believing to be true.

The basic thesis of the book is that, when faced with decisions or judgments we have two ways of thinking: 'System 1' and 'System 2' (not literally different systems, although that is a convenient way to talk about them). System 1 is the more instinctive, faster way to make decisions, often based on a very casual perusal of the evidence combined with gut feeling. System 2 is more considered and a slower attempt at rational evaluation. The bad news is that System 2 is hard work, so we instinctively tend to use System 1 and, secondly, that both systems tend to lead us to wrong answers.

An example of how System 1 gets things wrong is the 'Linda problem' – subjects in an experiment are told some details about a young woman who cares deeply about social problems and then asked to guess whether it is more likely that she is now 1) a bank teller or 2) a feminist bank teller. Most people give 2) as the answer in spite of the fact that basic logic would show us that all feminist bank tellers are bank tellers, so 1) must be the more probable answer. Effectively, System 1 has ignored the actual question and substituted an easier question ('Is it more likely that Linda is a feminist or a bank teller?'), rather than using System 2 to analyse the probability properly.

Another persistent cognitive bias about statistics comes in the 'Law of Small Numbers', which is our tendency to make sweeping judgments about a population based on a small sample. And this also holds in reverse: we expect a small section of a random group to reflect the statistical make-up of the whole. For instance, when asked to write down the results of an imaginary chain of six coin tosses (out of a chain of a hundred) most people will hypothesise a chain in which heads and tails come up equally often, whereas this will be less common than other combinations. Worryingly, even scientists who regularly use statistical methods are prone to errors of this sort and when asked to project the likelihood of the results of a small sample holding for the entire population, they will be far too confident.

One practical effect of this law is that we underestimate the role of luck and randomness in many situations. If, for instance, a manager has a succession of three good years, we will see this as proof that they will continue to be successful in the long-term, when it is far too small a sample for us to rule out the likelihood it was sheer luck. And this leads us to overestimate the role of skill in success (and failure) in general (a mistake that is sometimes known as 'survivor bias').

Kahneman writes brilliantly about these and many other cognitive biases: for instance, the anchoring effect (in which we tend to make an estimate close to the last figure mentioned), the availability bias (in which recent events affect our decision more than ones further in the past) and the focusing illusion (in which the thing that we are currently thinking about takes on a psychological importance out of proportion to its real importance). Occasionally, you might want to argue with the conclusions that the psychologists have drawn from a particular experiment, perhaps feeling that the subjects have essentially been tricked into looking more irrational than they are. On the whole, the evidence is inescapable – we make most of our decisions on the basis of fairly shaky guesswork and leaps of faith.

Reading about and recognising all of these biases can, in theory, help us to avoid making obvious mistakes in future (although it is clear from reading Kahneman how hard it is to apply this filter – he admits himself to many occasions on which he has made the very mistakes he has been studying for so long).

You may note as you go through this book how many other titles can be compared to this one. It could be argued that this is a key text and that once you have read it some of those other titles are dispensable. At the very least this would be highly recommended as a starting point for understanding how unsmart we can be in practise and to start trying to work out how to avoid the cognitive biases we are all prone to.

THE SPEED READ

Humans are not as good as we like to think at making judgments or coming to decisions. We are subject to the optimism bias, the anchoring effect, the availability bias, misunderstanding the Law of Small Numbers and many other cognitive biases. When we think with 'System 1', we make quick decisions based on our gut feeling and are often wrong. If we make an extra effort and use the more rational and slow 'System 2', we do a bit better but still make mistakes. Basically, people are really irrational and when we do something stupid, all we can do is try to slowly learn to be less stupid next time.

NB: A longer version of this review previously appeared in *A Brief Guide to Business Classics* (2017), by the same author.

The Willpower Instinct

Kelly McGonigal, 2011

A lot of people would identify their problem in life as being the result of a lack of willpower. If you overspend, engage in the compulsive use of drugs or alcohol, procrastination, not eating well enough or exercising enough, or have issues with anger, a lack of self-control is often at the heart of them.

There is an obvious niche for books that address this problem. Kelly McGonigal is a lecturer at Stanford University who started to give courses on the science of willpower. These led to this book, which is essentially a self-help book, but one deeply rooted in science. She writes, for instance, 'Science is discovering that self-control is a matter of physiology, not just psychology. It's a temporary state of both mind and body that gives you the strength and calm to override your impulses.' First, let's take a look at her approach.

She starts out by saying, very reasonably, that people who are in control of their desires, emotions and acts are generally better off than people who aren't. The best way to try to do better in this respect is to gain self-knowledge, to understand what is going on in situations in which you lose control.

Willpower has several aspects: 'I want', 'I won't' and 'I will'. What this means is that a desire can be something we need to resist (for instance, when we are drinking, horny, hungry or sleepy and feel the urge to give in to those feelings), or it can be something we need to say 'Yes' to: for instance, when we need to summon the willpower to perform a task.

The author discusses the separate parts of the mind that are responsible for our conflicting impulses, pointing out that willpower is situated in the prefrontal cortex, the part of the brain capable of complex planning and of moderating our social behaviour. The older, primal part of our brain is programmed to respond to satisfy our desires. 'Some neuroscientists go so far as to say that we have one brain but two minds – or even, two people living inside our mind. There's the version of us that acts on impulse and seeks immediate gratification, and the version of us that controls our impulses and delays gratification to protect our long-term goals.'

The book contains many exercises through which you can better come to know when your wiser self isn't in control. In particular, McGonigal recommends the use of meditation. As she says, even 'bad meditation' can help you to understand yourself as you can start to recognise those moments when you leave the calm, contemplative state in which willpower is feasible, and give in to those distractions that lead to temptation.

As you start to see how your habits and distractions put you in a state where your willpower is weaker, you can use various strategies to combat your worse self. One is to 'pause and plan': choose to delay the gratification of your desire by ten minutes and use the time to deploy your wiser self. A compulsive snacker might focus on the healthier self that will result from better choices. The prefrontal cortex isn't as quick as our primal instincts and giving ourselves a bit more time can result in better decisions. We also respond most strongly to immediate stimuli, and need to find ways to think of our future selves as though they are us right now. Try to remember that the future will be much like the present and your future self will be no better at sorting out the mess you are leaving behind for it to clear up than you are now. Do your (future) self a favour and don't leave that mess behind . . .

It is also important to cut the stimuli that lead to the loss of self-control. One thing to bear in mind is that if you focus too much on the thing you are trying to avoid, it can be a stimulus in itself. Thinking, I will avoid junk food, can increase cravings – think, I will eat healthily, which will be a more effective strategy.

There are some interesting scientific asides: for instance, a low-carb diet reduces fluctuations in blood sugar and this gives you more willpower over the course of time. However, a small dose of sugar can help in the short term if you are stressed or tired. Variability in heart rate is also important, as it covaries with willpower: one way to regain control is through meditation exercises that help you slow down your breathing. Also, it is interesting to read that congratulating yourself too much for short-term success in exercising will-power can weaken your resolve over time as you feel you've already attained your target.

Some of the weaker parts of the book come where it falls into the self-help cliché trap. McGonigal suggests that stress is the biggest enemy of willpower and her remedies include avoiding the news (because it will put you in a state of fear) and avoiding hanging out with other low-willpower people. But after this book came out McGonigal had a Pauline conversion on stress: she went into the research that shows it is your response to stress rather than the stress itself that can be harmful. This suggests that if you respond in the right way, being stressed can help you to be healthier, as she would go on to concede in her later book *The Upside of Stress*. This generous admission detracts from at least some parts of *The Willpower Instinct*. However, if you know that from the start, there is plenty here that could genuinely be a help to someone who is struggling with willpower.

THE SPEED READ

Willpower is central to our lives. We effectively have two minds: our primal instinct wants instant gratification (too easy to achieve in the modern world) and our wiser mind (the prefrontal cortex) allows us to plan for the future and to pay attention to wider issues. The trick to willpower is self-knowledge through mindfulness, allowing us to be aware of when and how our willpower tends to fail.

Mindset

Dr Carol Dweck, 2012

One of the more toxic elements of the US self-help genre is mindless reiteration of the idea that positive thinking is some kind of magic bullet.

Used in its least poisonous form, in a book like Peale's *The Power of Positive Thinking* (of which Donald Trump is a particular fan), this mainly comes with an emphasis on how positive thinking can help you to succeed and overcome failure. At its worst, in a book like Rhona Byrne's *The Secret*, it can degenerate into suggesting people who get ill or fail in life can blame themselves for just not having thought positively enough.

Given these pitfalls, it is a relief to find a book that focuses on that part of the positive-thinking idea that really does make sense: the proven fact that your attitude to life can make a real difference to how you learn and progress.

Dr Dweck is a psychologist who has extensively studied how our mindset affects our life. She distinguishes two kinds: a fixed mindset sees not only attributes and qualities as unchanging, but leads those who have it to avoid challenging themselves to try to grow or learn to do difficult things at which they might fail. A growth mindset reacts differently: it doesn't see failure or setbacks as bad things, but recognises the effort that has gone into a venture. This is more important than the outcome and leads to a person wanting to learn and develop, to hone their attributes and qualities. If you see abilities as attributes you can acquire, you are more likely to make the effort to do so.

There is a spectrum that runs from fixed to growth mindsets, which has two different aspects. You can have a fixed or growth mindset about different tasks: for instance, you may see physical strength as something that can be acquired, but not abilities such as being able to learn the piano. You may also adhere to the idea that other people are more trainable or adaptable than you. In fact, mindset is just a set of beliefs, regardless of the task or the specific person.

Dweck argues that the best outcomes are generally achieved if you adopt a growth mindset in every aspect of your life, including the romantic. One obstacle to doing this is the fact that we instinctively don't like to fail. It's simply easier to assume we can't improve in certain aspects and to choose easy tasks to avoid setbacks. Given this starting assumption, we fail to see that adopting a growth mindset is one way of overcoming the fear of failure. Yet it can convert a setback into another exciting step on the journey of progress.

The really interesting thing is that it is possible to influence how people see the world. If, for instance, a teacher tends to praise children for how smart they are (intelligence being a genuinely fixed quality), this will not inspire them to adopt a growth mindset as much as if the child is praised for their persistence and effort (which are not fixed and are growth mindset qualities). The same applies to everyone from employees to sportspeople to musicians. The more you encourage people to value learning and effort over easy success and praise for fixed qualities, the more you foster the growth mindset.

This is all easy to say, but Dweck roots it in some fascinating research. In one of her best-known studies, she first gave a group of kids an easy jigsaw to complete and then offered them the chance to do the same puzzle again, or try a new, harder one. They split into two groups: with the fixed mindset opting for the easy task.

The really interesting part came when she studied the brainwaves of the two sets. The kids with a fixed mindset were only interested in those parts of the feedback they were given that reflected their current level of ability and tuned out anything that might help them learn to improve, not even showing interest in hearing the right answer to a question they had got wrong in a quiz. They had already labelled their performance as a failure and that meant that hearing the right answer held no appeal.

The effect of feedback on mindset was demonstrated in even starker terms in an experiment in which two groups of kids were given a non-verbal IQ test. In one group the kids were praised for being intelligent when they did well, while in the other they were praised for having tried really hard.

When the two groups were offered the chance to try a harder test, far more of those who were praised for their efforts were willing to try. This included some of those who had initially tested as having a fixed mindset, directly demonstrating the effects of a different teaching style. However, if the kids who had initially shown a fixed mindset did badly on the second test, they fell back into the binary trap of success/failure, judging themselves in harsher terms for not doing well than the kids who already had a growth mindset.

There are real messages that can be taken from this book, both about how we treat other people and how we think about ourselves. The key is to always remember we can get better, and not to care so much about being praised for success. As Dweck says, 'Why waste time proving over and over how great you are, when you could be getting better? Why hide deficiencies instead of overcoming them? Why look for friends or partners who will just shore up your self-esteem instead of ones who will also challenge you to grow? And why seek out the tried and true, instead of experiences that will stretch you?'

---THE SPEED READ---

People tend to have either a fixed mindset (seeing their abilities as fixed and avoiding failure or challenges) or a growth mindset (seeing their abilities as adaptable and treating failure as part of a learning curve). We can be trained or train ourselves to have more of a growth mindset in every aspect of our life, and there are huge benefits to doing so.

The Signal and the Noise

Nate Silver, 2012

N ate Silver is a statistician, best known for making some brilliant political predictions, in particular his near perfect prediction of the 2012 US election – calling Obama's victory state by state. That success led to this book becoming an overnight bestseller.

Silver made his reputation with his analysis of baseball statistics. He developed PECOTA (Player Empirical Comparison and Optimisation Test Algorithm), a tool for predicting players' performances. The model incorporated some novel features such as tagging players against 'comparable' players to get a clearer picture of their differential value and using unique sources of information, including data from a radar gun that measured the speed of the ball as it was pitched.

In *The Signal and the Noise*, Silver discusses the use of statistical models to make predictions about such things as politics, investment, sport, and hurricanes. In pursuit of excellence, he visits and talks to some of the most successful practitioners in these different fields, trying to understand the best ways of separating out the 'signal' from the 'noise'.

The well-known phrase 'correlation does not imply causation' means two seemingly related variables may be separate. Silver gives as an example, 'Ice cream sales and forest fires are correlated because both occur more often in the summer heat. But there is no causation; you don't light a patch of the Montana brush on fire when you buy a pint of Häagen-Dazs.' The noise is the variables that are uncorrelated to the event we are trying to predict, while the signal is, in essence, the variables that do have a causal relationship. The more our model incorporates the latter and excludes the former, the closer our prediction will be to the actual result.

As humans, we are highly prone to seeking patterns and often think we see one where there is none. This is the source of many problems for forecasters. One thing Silver observes across all the fields he investigates is the amount of humility shown by the best forecasters, who are always willing to explore and examine flaws in their models and to recognise their limits.

Silver has always argued that it is important to give results as a range of probabilities rather than as a straight projection. For instance, rather than saying 'Trump has the edge in North Carolina,' he might say 'Trump has a 61 per cent chance of victory in North Carolina.' This reduces the danger of the listener having excessive confidence in the prediction. (It's interesting to note that Silver's 2016 prediction of a 30 per cent probability of Trump winning is often cited as a failure by him, whereas the result depended on variation in key states that Silver had been quite humble and cautious about.)

This also leads to the 'prediction paradox': the less confidence we have in our predictions, the more likely we are to get them right. The forecasters Silver talks to always doubt their results and look for ways in which their models might be tweaked. He describes a group of people who are always assiduous, hard-working and imaginative about how their current solutions may be flawed.

Silver's book has some crucial morals for those wanting to improve their thinking. He believes we are generally torn between various urges. We want definite, straightforward predictions that we can act on like, 'It will rain today'. This leads us to look for methods and techniques that are guaranteed to succeed. And we tend to stubbornly stick to our predictions, even as it becomes clear that the facts have changed. The best forecasters avoid all these pitfalls, opting for clear statements of probability, complex methodologies, and humility about failure.

This is a moral that can be taken in many areas. No matter how certain we are that our current thinking on a situation is accurate and correct, we need always to have the humility to consider the possibility that we could be wrong, and to examine the reasons why we might be wrong. Failure to do so is always liable to leave us stubbornly clinging on to a false or unhelpful belief. Being flexible and humble when it comes to confidence in our beliefs is a crucial way to avoid errors.

---THE SPEED READ---

'Distinguishing the signal from the noise requires both scientific knowledge and self-knowledge: the serenity to accept the things we cannot predict, the courage to predict the things we can, and the wisdom to know the difference.' A clear-minded guide to the art of prediction.

Train Your Brain for Success:
Read Smarter, Remember More, and Break Your Own Records

Roger Seip, 2012

Seip starts his book with an anecdote about how, when he was nineteen, he went to see the motivational speaker Mort Utley, who began his talk by pointing out that most people don't get what they want in life. Seip's initial response was to think that this was the most un-motivational, depressing thing he had ever heard. But then he had an about-turn, realising that what Utley had said not only made sense but that that was the precise reason why Seip was there in the first place.

This kind of gung-ho anecdote immediately raises the hackles, suggesting this is going to be a motivational, happy-clappy sort of read. And it doesn't help that Seip goes on to suggest that the reason most people aren't successful (which he seems to equate with being slim and having a relationship and a high income) is that 'a big part of you wants just to be average'. He suggests our desire to 'fit in' is at fault here: fitting in with the ordinary, inadequate majority is in fact a path to 'failure'.

There is a tendency in this area of writing to blur the boundary between smart thinking and success and this book certainly falls into that category. Seip gives the reader a pep talk about how easy it is to be successful. You just need to learn the fundamental elements and make them into habits if you want to rise above the unwashed herd. He recommends the reader should seek out people who have succeeded in the same area they would like to succeed in and learn to think the way that they do. Trying to acquire the same thought patterns and habits can help your own quest. So far, so 'positive thinking': we have a reticular activating system (RAS) in our brain, which allows us to focus while filtering out the things we want to ignore. Visualisations can help with this, as the 'more clearly you can picture the results you want, the more likely you are to deliver the performance that produces those results.'

The book also has some very practical advice on ways of improving your memory using filing systems and mental tricks. The key here is organising your memories. The organising system of the brain is referred to as the 'mental file folder system', or 'file image glue' (FIG for short). The glue is the element that makes a memory file stick and the best way to do this is by using images that carry an emotional charge or a cue for an activity. In essence, if you want to remember a shopping list of oranges, pilchards, and cupcakes, then remember a time when cupcakes made you happy, while imagining squeezing an orange under your armpit while bathing in pilchards. OK, it doesn't have to be that vile, but you get the idea.

There's an interesting chapter on why you 'read like a sixth-grader' and how to improve your reading. He points out that an inability to focus on what you are reading means your brain isn't being stimulated enough and you should read faster to combat this, absorbing more information in less time. He provides some useful tools for warming up the brain (supplemented by further exercises that are accessible online).

It's not all happy-clappy. And there are also some images which are at times quite funny: for instance, who couldn't love a Venn diagram with two circles, a large one labelled 'where the magic happens' and a much smaller, entirely separate one – 'your comfort zone'. However, there are better books out there on the most important areas, such as memory and information absorption, and the gee-whizz tone of the book may be enough to put many readers off.

THE SPEED READ

Look around you at the lumpenproletariat. Mediocre, fat, underpaid, single losers everywhere. Do you want to be like them? You say you don't, but you do really, don't you? If you really don't, then in order to be successful, thin, rich and happily married, get out of your comfort zone, go where the magic happens, and learn the thinking habits that can help you on the way.

The Undefeated Mind:
On the Science of Constructing an Indestructible Self

Dr Alex Lickerman, 2012

It can be hardest to think clearly when you are going through adversity or it has recently affected you and you are struggling to recover. In response, this is a book by a physician with a strong Nichiren Buddhist streak. It is about how to overcome suffering, whether it comes in the form of a job loss, serious illness or bereavement. Part of the key to this is the ability to derive wisdom from adversity.

Lickerman uses a mixture of science and anecdotes to focus our attention on the things that genuinely make or break us. Through this we can develop the resilience needed to deal with life's toughest challenges. At times he takes a rather Stoic tone, urging the reader to develop a mind that can face any combination of circumstances with calm and even happiness.

A large part of the book focuses on happiness, and this review will concentrate more on the smart-thinking aspects.

Lickerman looks at the problem of being discouraged when we face failure. He points out that we need a wider mission to develop a sense of endurance: individual strategies may fail, but the mission will still be there if we take this approach. A personal mission reduces the temptation to quit after a setback.

He also suggests activities – whether it be overcoming pain, or achieving a task – for a greater meaning. You shouldn't look for an exciting mission, just one that gives your life a purpose which, in itself, will make the strategies you use to achieve it more exciting. 'I don't know if even a sculptor gets excited about filling the world with beauty. He gets excited about sculpting – but only because that's how he fills the world with beauty, the activity that makes his life feel most significant.'

This leads the practitioner to occupying a state of mind in which they learn to look for lessons and positives from negative events and setbacks. 'From a Buddhist perspective, however, this means neither denying our problems exist nor denying they make us suffer . . . for when confronted by harsh circumstances over which we have no control, we become capable of enduring them only by finding a way to create value with them.'

If, along the way, you find you have lost self-confidence, and are dithering over a problem or challenge, Lickerman suggests jumping into action, on the basis that you will regain self-confidence and more perspective on the scale of the problem through trial and error than you would from just ruminating and dwelling on it.

Part of his point here is that, after a setback, people can procrastinate on the basis that they need to recover the feeling before they take action. For instance, someone who has come out of a failed relationship may dither, thinking they need to wait until they feel happy before taking action, whereas the mere act of taking action may be the first step towards regaining the happiness they have lost.

There is a similar loop in place when it comes to lacking self-confidence: it can lead to having negative beliefs, and that feedback further dents the self-confidence. It's only by taking action and banishing those negative beliefs that we are likely to regain the self-confidence.

The religious elements of the book can be jarring if you aren't a fan of Nichiren Buddhism. Lickerman makes many excellent points, but often suggests that his approach or advice is very specifically Buddhist, to the point that it feels like he believes that only his strand of the religion has the solution for adversity. But if you can get over that drawback, this is a useful guide to the difficult art of thinking clearly in difficult times, or regaining our ability to think and act after experiencing challenging events.

THE SPEED READ

If you are suffering from a lack of confidence and negative thinking following a serious setback in life, then you need to find a way to treat that adversity in wisdom with calm and equanimity. Take action without waiting for yourself to 'feel' better first, as feeling will follow action. And develop a personal mission that will give meaning to your life and will make your strategies for fulfilling that mission more exciting.

Big Data:
A Revolution That Will Transform How We Live, Work, and Think

Viktor Mayer-Schönberger and Kenneth Cukier, 2013

One thing that has been revealed in recent years is that large corporations and governments gather a huge amount of data about us: from Facebook's use of our images to Google analytics, and from the local council's records of our payments to the detailed records collected by state agencies such as the US National Security Agency (NSA).

Our emails, video chats, holiday snaps, official documents and our website logins are all out there in one form or another, and many of them feed into the algorithms used in seemingly innocuous ways (such as when companies target appropriate demographics through social media) as well as in more sinister fashion (for instance, in the way that information harvested by Cambridge Analytica from Facebook profiles was being used to target political advertising, potentially influencing elections around the world).

Big Data, by Viktor Mayer-Schönberger and Kenneth Cukier, is an extended exploration of the implications and significance of the flood of information that is being channelled in so many ways around us. The term is shorthand for the way in which huge amounts of data are crunched and analysed to give crucial information about anything from what colour of used car is most likely to be in good condition, to identifying the most dangerous manholes in New York City. The results may often be counter-intuitive or surprising: the authors suggest that this is evidence of how powerful big data is, since it can reveal trends and phenomena we wouldn't otherwise recognise.

Their basis thesis runs counter to the traditional scientific method of using correlation to give us evidence to prove or disprove a theory about causation. Standard statistical methods rely on gathering enough evidence to rule out the possibility of bias in our sample. The authors argue, contrary to this, that now that we have an unlimited supply of data at our fingertips, we can prove trends and phenomena are real even if we don't understand the causal process that leads to them. They also suggest that the quality of individual pieces of data and the possibility that they have been measured inaccurately doesn't matter so long as there isn't a systematic bias in the method used to gather the data: the sheer quantity of the data will overcome the need for such niceties, and governments and corporations can make reliable plans and predictions based on big data even in cases where there is no understanding of the correlation or causality.

There is some evidence for this working. The book *Moneyball* by Michael Lewis (see p.79) shows how Billy Beane, general manager of the Oakland A's baseball team, used statistical analysis to hugely improve the performance of his team. Walmart crunched the numbers to find that sales of snacks like Pop-Tarts tended to spike at the same time as emergency supplies like flashlights when a big storm was on the way. They stocked Pop-Tarts next to the hurricane supplies in storm season and boosted sales yet further.

The book relies heavily on the authors' interpretations of examples from previous popular science books such as *Freakonomics* (see p.97) and *Thinking, Fast and Slow* by Daniel Kahneman (see p. 144), rather than on original research and examples. It is also dependent on a big dollop of hype about the many ways in which big data is about to completely transform the world, which can be quite persuasive on first reading. It's easy to let the authors persuade you that in the future, terrorists will be caught before they even start planning their first attack, and all your needs and wants will be met or soothed by algorithms that know what you want better than you do.

The trouble is that there is plenty of evidence of the ways that big data can go wrong. The authors big up the Google Flu Trends project, in which global searches for flu-like symptoms were being used to predict outbreaks of influenza, claiming that as it predicted the H1N1 (swine flu) outbreak in 2009, it was now more reliable than other means by which governments might make such predictions. GFT proved to be far less reliable than the authors assumed it would be and analysis showed that it was significantly wrong in a large percentage of cases. This was partly because the terms used by Google as predictors could equally apply to too many other illnesses, and this led to over-prediction, and a significant degree of inaccuracy.

There is a real danger here of over-confidence in big data. To be fair, the authors do recognise some of the other pitfalls of trusting it: they engage in a lengthy discussion of the implications for privacy in the future, noting that in the digital world, '[privacy] laws constitute a largely useless Maginot Line. People willingly share information online – a central feature of the sendees, not a vulnerability to prevent.'

They also discuss the scary topic of 'predictive policing', in which data is used 'to select what streets, groups and individuals to subject to extra scrutiny, simply because an algorithm pointed to them as more likely to commit crime'. The obvious risk is that computers may declare us to be guilty until proven innocent.

So, while the authors spend much of the book hyping the potential and achievements of big data, they do at least end up by warning about the dangers of over-reliance on this method of analysis.

┌─**THE SPEED READ**────────────────────────┐

A guide to the world of big data, in which the sheer quantity of information gathered by corporations and states means that there is no longer any need to prove causality when it comes to predictions and strategies: we can rely on correlation alone to discover amazing trends and phenomena. The problem is that the evidence suggests that big data is no more reliable than traditional statistical methods, and can lead to overconfidence in inaccurate predictions.

Focus:
The Hidden Driver
of Excellence

Daniel Goleman, 2013

Life is a sprawling, complex, vast thing that can't really be reduced to simplistic summaries. This is a problem for publishers, as the titles that tend to sell best are snappy, attention-grabbing and short. This often leads to books making excessive claims beyond the scope of the quite narrow subject being covered.

Focus is an example of such a book. That hubristic claim in the subtitle isn't really justified by the book's contents, interesting read though it is in places. Goleman – the author of *Emotional Intelligence*, among other titles – is a convivial guide, and there are certainly plenty of interesting titbits here. The basic theme is that the modern world, with its flood of social media, screens and messaging, is making it harder for us to focus on important tasks.

The argument is that we are prone to suffering from cognitive exhaustion after a period of concentration, and this is only being exacerbated by the digital world in which we live. This also affects our interpersonal relations, as we are rarely in the present moment. 'No birthday, concert, hangout session or party can be enjoyed without taking the time to distance yourself from what you are doing, to make sure that those in your digital world know instantly how much fun you are having.'

Goleman distinguishes between 'bottom-up' involuntary mental processes and 'top-down' intentional processes that require focus. He also talks of three different kinds of focus, applied by truly successful people. Inner focus is based on knowing what is going on in our own minds, understanding our values, dealing well with distressing emotions and maintaining motivation. 'Other focus' is our ability to pay close attention to other people, which is the root of all interpersonal skills, including persuasion, negotiation and communication. 'Outer focus' is about how well we perceive the larger forces in our world, whether it be the dynamics of our workplace or the ways in which economic and environmental forces are changing the world, and how that will affect us.

You may have noticed that it is a bit of a stretch to call 'outer focus' a type of 'focus'. The trouble is Goleman has to do this to support his wider thesis and the idea that a true leader demonstrates 'triple-focus'. The latter point is illustrated by examples such as Steve Jobs who narrowed Apple's focus to four main projects and put energy into spotting future trends.

Goleman is always fascinated by the workings of our minds, and occasionally gets a bit geeky in his explanations. For instance, he writes that 'tightly focused attention gets fatigued – much like an overworked muscle – when we push to the point of cognitive exhaustion. The signs of mental fatigue, such as a drop in effectiveness and a rise in distractedness and irritability, signify that the mental effort needed to sustain focus has depleted the glucose that feeds neural energy.'

However, he is more interesting when it comes to practical advice on how to deal with cognitive exhaustion and refocus. For instance, he talks about that moment when you keep staring at the screen and find yourself switching off. His advice is to consciously take a break and to let your mind wander. Often you will find that after a brief break doing this, your mind wanders back to the task at hand, and often will find more creative ways of solving any problems you are facing.

He also talks about how important it is to work on things that you love, and how much easier it is to focus on a task when you love it. Basically, when you are doing something you enjoy, it doesn't seem like an effort and your brain tires more slowly. Finally, he suggests that when the impact of a problem in your life is distant, you should imagine it as being something which is a problem in the here and now. For instance, if you have a fence that hasn't been weather-protected, you should picture it falling down after it has rotted away, and this will motivate you to focus on it now, rather than to procrastinate.

This latter point may seem slightly elaborate, but again Goleman is laying the foundations of a higher aim. He ends the book by suggesting that the reason we are failing to get to grips with the scale of the climate change crisis is because we have a cognitive bias towards current problems. Effectively, we don't have enough 'bandwidth' to deal with future threats. We need a dose of that 'outer focus' we mentioned earlier and the ability to focus on future problems as though their impact is imminent.

This brings the book to a worthy and admirable conclusion, but leaves the reader feeling that those chunks of the book that seemed a bit tendentious at the time were all set-ups for this final act. This unfortunately provokes the sense that, enjoyable as it was spending some time in the company of Mr Goleman, the book ends up being, well, unfocussed . . .

THE SPEED READ

You are bombarded by emails, social media and messages, and this is exacerbating the problem of cognitive exhaustion. To regain focus, take a break, let your mind wander and make sure you are working on something you love if that is possible. A visionary leader needs to focus on themselves, other people and the bigger picture. And if people only listened to a visionary like me, solving the problem of climate change would be a no-brainer.

Intuition Pumps and
Other Tools for Thinking
Daniel C. Dennett, 2013

The basic pitch of this book is it contains seventy-something 'tools for thinking', which should make it the ideal candidate for a collection of smart-thinking books. It is, in some ways, but possibly not in the way that the pitch suggests.

Dennett is a brilliant philosopher and polemicist, who has most notably written extensively about evolution, consciousness and free will in books such as *Consciousness Explained*, *Darwin's Dangerous Idea* and *Freedom Evolves*.

He begins *Intuition Pumps* . . . with twelve traditional tools for thinking, including Occam's razor (in which we assume that the simplest explanation for something is probably the right one), Sturgeon's law (suggesting that 'ninety per cent of anything is crap') and the use of rhetorical questions. He points out that the use of 'surely' in any argument almost always reveals a lack of logical foundations. He discusses the use of 'jootsing' (a term coined by Douglas Hofstadter to mean 'jumping out of the system'), and *reductio ad absurdum*. And, interestingly, he gives his own version of Rapoport's rules, intended to keep any argument or debate civil and reasonable.

This includes summarising the other person's arguments so fairly that they say, 'Thanks, I wish I'd thought of putting it that way', listing points of agreement, acknowledging anything you've learned from the other person, and doing all this before saying a word of criticism. We'll come back to that later.

Most of the rest of the book is taken up with summaries of Dennett's thoughts on topics he has written about elsewhere, and a description of the ways in which we can use those intuition pumps – basically, methods of thinking that allow us to say, 'Of course, now I see it is obvious' – to explore these topics.

We have, for instance, a discussion of John Searle's Chinese room: this is a traditional thought experiment in which Searle imagined sitting in a room with two slots. From one slot he receives English messages. He uses an automatic device to translate these into Chinese, which he doesn't understand, and passes them out the second slot. The experiment is designed to show the irreducibility of consciousness and the impossibility of a machine being conscious, since a machine can appear conscious while similarly just following a set of rules. Dennett cleverly rebuts the argument, pointing out that we are being tricked into focusing on the consciousness of the person in the room, whereas any actual consciousness (if it were to exist) would be manifested at a wider level of the entire system, including the room and method used for the inputs and outputs. This would need many other attributes other than the limited ones described, so the description of the room's set-up is effectively a disingenuous one.

On the subject of consciousness, Dennett discusses at length the idea of philosophical zombies – beings who are just like humans that experience nothing. Similarly, the species of wasp called Sphex demonstrates behaviour that can appear to be entirely mechanistic and non-intentional, when they repeatedly drop their prey outside a nest and inspect inside, repeating these same motions as though stuck in a loop: they are effectively behaving as though they are automatons at these times. Dennett draws out all the possible ways of thinking about these examples in order to show all their implications for free will, evolution and consciousness.

In general, Dennett is brilliant at engaging with real science (something many philosophers avoid) and his arguments are often compelling. He is good at making consciousness seem far less mysterious than it does in many other accounts and at explaining the ways in which the phenomenon of consciousness arises from a purely material world. He argues that there is no 'hard problem' of consciousness that undermines evolutionary, materialist explanations of the mind, and that those who say that there is are using misdirection.

He treats the term 'free will' as something that mostly tends to obfuscate arguments about the human condition but none-theless attempts to give a clear account of its limits. For instance, free will can't mean 'changing the future' from 'what was going to happen' to 'what is going to happen', since that is incoherent.

Dennett can also be pretty funny. He writes, for instance, 'I think we should stop treating "God works in mysterious ways" as any kind of wisdom and recognise it as the transparently defensive propaganda that it is. A positive response might be, "Oh good! I love a mystery. Let's see if we can solve this one, too. Do you have any ideas?"'

Dennett concludes with some thoughts on why philosophy is still valuable in the modern world, and large parts of the book vindicate his assertion. He provides a series of models for considering an argument or statement and exploring its foundations. One can ask, 'Are there similar arguments that I can compare to this?', 'What would happen if I changed part of this argument?' and many similar questions designed to tease out the underlying assumptions that are being unconsciously made. A good example is his treatment of 'wonder tissue', arguments that are based on supernatural assumptions.

The caveat is that Dennett can also be a combative grinch of a writer, who seems to go out of his way to revisit old feuds: for instance, he often clashed with the late evolutionary biologist Stephen Jay Gould. He introduces three types of 'Goulding', which he calls 'rathering', 'piling on' and 'the Gould two-step', which he implies are deliberately deceptive. This really doesn't seem to fit in with the spirit of Rapoport's rules and can leave a bit of a bad taste in the mouth.

The best way to read this book is not as a quick way to learn some new thinking tools but as an overview of the work of one of the most interesting contemporary philosophers. As it covers ground from his other major works in a more concise fashion, it could even work as an introduction to his thinking, although there is a danger that a reader new to Dennett might be put off by his occasional slip from following his own rules of etiquette.

THE SPEED READ

An intuition pump is a tool to use in an argument or statement with which you can take a fresh approach. You can make clear what assumptions the argument rests on and whether the same kind of argument would be valid in a different but related situation. The traditional thought experiments of philosophy are thus intuition pumps, as are ideas such as Occam's razor. In this work a brilliant contemporary philosopher gives an overview of his views on evolution, consciousness and free will and explains why he is always right about everything.

Seeing What Others Don't

Gary Klein, 2013

Seeing What Others Don't takes a fresh approach to the whole idea of smart thinking and creativity, focusing exclusively on 'insight'. Klein defines insight in thinkers thus: 'Intuition is the use of patterns they've already learned, whereas insight is the discovery of new patterns.'

He gives the example of two cops in a car behind a fancy BMW. The younger cop says, 'Did you see that? He just ashed his car.' He had noticed the driver discarding a cigarette on the upholstery of the car. The officer was immediately suspicious at the sight of someone apparently treating their own property so badly and the older cop saw the insight as a piece of genius. They immediately put on the lights and the car took flight – it had indeed been stolen. It was the younger cop's ability to notice the unusual that had unlocked the story unfolding in front of their eyes.

It is the 'unexpected shift in the way we understand things' that is Klein's focus. And, of course, a shift in understanding can have a wide impact on emotions, behaviours and goals, so it is more than just a different view: it can change our interactions with the world. In business, an insight is something that changes the way you understand the market you are supplying, for instance. But in any field you can find a moment when the paradigm shifts and you see things in a new light. One really basic example comes from the work of the philosopher Ludwig Wittgenstein: he drew a simple picture that looked like a duck when viewed in one way, but could also be a rabbit. The brain finds it hard to see both interpretations at the same time and switches between the two versions.

In addition, one whole strand of the philosophy of science is based on this idea. The American philosopher Thomas Kuhn suggested that scientists mostly accept the paradigm they first learn. When their results don't quite fit it, they will try to explain this away as an inaccuracy in the measurements, or an inconclusive bit of data. Finally, one scientist, somewhere, will suggest a completely new angle, whether it be the idea that the Earth actually rotates around the sun or the theory of relativity. Appropriately enough, Klein discusses the insights of Einstein, Darwin and Fleming in the book. But he also spends time exploring other areas in which insight may be useful, such as training firefighters, artists and in business.

For instance, he discusses the way that Wall Street expert Harry Markopolos was one of the earliest people to call out Bernie Madoff as a fraud, years before Madoff was unmasked as the man behind the largest Ponzi scheme in history. Markopolos was a somewhat eccentric figure who was asked to design a financial product that would rival Madoff's success. He concluded that there was no way of doing this and decided Madoff must be doing something illicit. But his early warnings were ignored at a time when Madoff was widely accepted and respected in the industry. Of course, now we know that the paradigm did indeed shift in the end.

So how do we learn to have more insights? Klein's three basic suggestions are that we 'look for connections', 'look for contradictions' and 'look for creative desperation'. Looking for connections is a way to understand your current problem through comparisons with a different area. The Japanese planned their attack on Pearl Harbor not through an exhaustive study of the location, but through a comparison with a Mediterranean battle that the Italians had fought with the British.

Looking for contradictions involves spotting things in a situation that simply can't both be true. In the late 1990s, hedge fund managers John Paulson and Paolo Pellegrini concluded that the widespread belief that had developed among financial commentators that it was feasible for house prices to rise indefinitely simply couldn't be reconciled with basic economics. They spotted the financial meltdown coming.

Finally, studying creative desperation is a way of trying to acquire the mindset that sometimes allows people or businesses in extreme situations to make an extraordinary turnaround, whether it be firefighters who are trapped and start another fire to open up an escape route, or companies who completely revamp their business plan and products to escape bankruptcy. In these situations, people or companies lose the fear of making a mistake simply by acting because they no longer have a choice. This makes them willing to take risks that they might otherwise not have done.

In addition, Klein considers the role of coincidences (as in the case of the doctors who were the first, before the identification of HIV, to notice how many gay men were developing similar symptoms) and basic curiosity (the way that penicillin pioneer Alexander Fleming noticed bacteria dying near a mould and wondered about the cause). At the same time, he warns that the three suggestions discussed earlier were responsible for a much larger proportion of the insights he studied and analysed for the book.

As with any book focused on a narrow subject, there is a degree of repetition, and the author can occasionally come across as self-congratulatory. But on the whole, by broadening the idea of creativity and viewing insight as the key break-through that allows for a paradigm shift, Klein's book is a useful and interesting addition to the smart-thinking canon.

THE SPEED READ

Insights are unexpected shifts in the way we see things that can lead to radical changes in our behaviour and actions. To increase your ability to have insights, you should focus on looking for connections between disparate situations and contradictions within a situation, and you should study situations in which people or companies that are in danger have taken desperate risks that have transformed their survival prospects.

Zig Zag:
The Surprising Path to Greater Creativity

Keith Sawyer, 2013

Keith Sawyer is a creativity researcher but, rather than writing about his research, he has chosen to write a practical manual outlining the basic steps of all creative processes.

He warns against making two common mistakes – thinking you don't have to constantly exercise your creativity and thinking there is 'one great idea' out there waiting for you. He then proposes eight steps to creativity, each summarised in a single word: ask, learn, look, play, think, fuse, choose, and make. This deceptively simple series is backed up with a variety of exercises and prompts that makes this, at the very least, a fun book to explore when thinking about the creative process. The text includes many business anecdotes, including a few old chestnuts . . .

For instance, he starts with 'ask' (ask the right question), and he tells the story of Howard Schultz, who drank a beautiful cup of Italian coffee in Italy and initially asked himself how he could set up an Italian-style coffee bar back in Seattle where he was working for Starbucks. They were then a wholesaler and provided seed money. His first attempt was faithful to the atmosphere and décor of his Italian experience, but he had only modest success, and went on to ask a different question. 'How can I create a comfortable, relaxing environment in which to enjoy great coffee?' This led onto a successful coffee chain – and he was eventually able to buy out Starbucks entirely and absorb them into his business.

The creative process often starts with finding the right question. It's important to note at this stage that Sawyer deliberately avoids depicting a linear process by which all of these stages have to be reached. He contrasts a linear diagram of the steps with one where you zig zag between them, reminding the reader that each step might lead you back to others.

Anyhow, the 'next' step is to learn: he points out that creative people are often voracious learners who never stop trying to find out more about different things.

'Look' refers to the heightened curiosity you need to truly be aware of the world around you and the curiosity that can lead you in unexpected directions. The more different ways in which you look at the world and understand it, the more likely you are to have a creative idea in it.

'Play' is self-explanatory: all creative processes involve an element of spontaneity, but again, Sawyer has some fun exercises, and makes some interesting suggestions. He asks you to consciously leave some of your work unfinished overnight so your subconscious can work on the problem.

'Think' refers to the process of constantly turning ideas and challenging yourself to ponder every single option. Creativity is, in a sense, a numbers game. The more ideas you work your way through, the more likely it is that something will spark. And you will often have to 'fuse' elements suggested by different ideas to come up with something truly original.

Finally, you will have to 'make' your idea a reality, by going through the hard slog of converting an ethereal concept into something tangible. And 'choose' which ideas you are going to take forward, since it is impossible to focus on too many things at once.

These kinds of atomised approaches to creativity are often unsatisfactory if they don't fully capture the messiness of the process. However, by acknowledging that real creative processes often involve zig-zagging your way through these steps and also by including plenty of smaller suggestions, Sawyer has avoided the worst traps of his chosen approach and produced a book that many readers will find inspiring.

THE SPEED READ

Creativity is ongoing, and you can't count on having one big idea. To generate worthwhile ideas, use eight steps: ask, learn, look, play, think, fuse, choose, and make. By zig-zagging your way through these steps, you can improve and better understand your creative flow.

Curious:
The Desire to Know and
Why Your Future Depends on It

Ian Leslie, 2014

The essential thesis of this book is the idea that, while
everyone has curiosity in their youth, some of us lose it as
we grow older. One reason why this is interesting is that it
seems related to the idea of growth mindsets and fixed mindsets
(see p.152). The idea that your abilities are fixed and immutable
seems intricately connected to a lack of curiosity about how
things could be different.

Ian Leslie draws on examples from psychology, business
and history to demonstrate the crucial importance of curiosity,
although he gets bogged down too often in the idea that the
internet damages curiosity. Information is too easy to access
online and forget, he suggests, and resources such as Wikipedia
don't foster the deeper connections between subjects that
create true learning and understanding. This comes across as
luddite and fogeyish at times. But in other places, his stress on
the idea that curiosity is at the root of discovery, creativity and
culture is mostly persuasive.

Leslie depicts curiosity as one of the basic human instincts driving our progress noting, for instance, that we appear to be the only creatures who look up at the stars and wonder what they are. (He might have thought twice about writing that had he read Daniel Gilbert's account of the pitfalls of such sentences: see p.101). Leslie suggests that, while curiosity can lead an animal into risky territory, it also aids expansion and survival, as various individuals dare to wonder what lies beyond the forest, the nearest mountain range, on the other side of the river and even beyond the oceans. He writes: 'What makes us so adaptable? In one word, culture – our ability to learn from others, to copy, imitate, share and improve.'

He notes that while the Chinese were once an advanced culture, ahead of the west, progress stagnated around the same time the emperors decided not to pursue further exploration. Meanwhile, European cultures were setting out to discover the 'new world'.

In acknowledging that curiosity isn't always a positive force, Leslie distinguishes between 'diversive curiosity' and 'epistemic curiosity'. The former is the pursuit of short-term novelty – gossip or puzzle-solving. By contrast, epistemic curiosity is the search for real comprehension and truth, the attempt to solve mysteries.

The author warns that the 'cognitive elite' are of increasing importance. 'The truly curious will be increasingly in demand. Employers are looking for people who can do more than follow procedures competently or respond to requests; who have a strong intrinsic desire to learn, solve problems and ask penetrating questions.' He notes that such individuals are not easy to manage, as they can be capricious, but that they are crucial to corporations.

He suggests approaching curiosity as a kind of mental

'muscle', one we have to flex to keep strong. He describes curiosity as a recursive process: every attempt to answer its questions leads on to more mysteries, more questions, and in the process that precious, childlike curiosity is rebuilt. He also highlights the importance of exploring different areas of knowledge. 'We know that new ideas often come from the cross-fertilisation of different fields, occurring in the mind of a widely knowledgeable person.'

Books that attempt to explore one fairly narrow human attribute, based on a single-word title, sometimes end up overstretching their subject and labouring the point. Leslie does include some fairly obvious stories of creative curiosity, with subjects such as Benjamin Franklin, Walt Disney and Steve Jobs. He also gets mired in psychological tests such as a version of the 'marshmallow test' – in which children are told not to turn around to look at a lovely toy. (Almost all children of a certain age fail this test, and this is suggested as evidence of the innate curiosity that we all have from birth.)

However, curiosity is a subject that suits quite a meandering approach, in which the author explores ideas such as how mutual curiosity affects a relationship, and even the role of curiosity in political discourse. 'Ignorance as a deliberate choice can be used to reinforce prejudice and discrimination.' We have Leslie's own (ahem) curiosity to thank for the fact that this rambling adds up to an enjoyable and intriguing journey.

THE SPEED READ

Curiosity is a fundamental human impulse that has driven our culture and evolution. We are all born with it, but some lose it with age. It is so important that we should work on honing or reviving our own curiosity, not in a shallow, distracted kind of way, but in a deeper way (going beyond the quick internet search) to try and really understand the world around us.

How We Learn:
The Surprising Truth About When,
Where and Why it Happens
Benedict Carey, 2014

Any book that promises to 'free your inner slacker' has an immediate appeal to plenty of readers. Who wouldn't want to be told to make less effort, sleep more and take more time out to play and waste time? This is particularly true of students struggling with intensive learning routines.

Science journalist Benedict Carey mixes psychology with pop science to take a fun approach to tackling the so-called 'enemies of learning'. His basic thesis is that distraction, interruption, forgetfulness, restlessness and even pure laziness can actually be virtues when it comes to the process of learning. For instance, he argues that if you often forget the things you are trying to learn, this can be part of a positive process in which your brain is effectively filtering information and gradually whittling down the things it really needs. When people in spelling bees finally fail, often stumbling on a word they do know, it is because they are remembering too much.

'If recollecting is just that – a re-collection of perceptions, facts, and ideas scattered in intertwining neural networks in the dark storm of the brain – then forgetting acts to block the background noise, the static, so that the right signals stand out. The sharpness of the one depends on the strength of the other.'

There is some evidence for this, such as the 'forget to learn' theory – the idea that memory is strengthened by the act of forgetting something then being forced to retrieve it. He also makes a case for the virtues of being interrupted: people remember more if they spread their studying time out (the 'spacing effect' or 'distributed learning'). Having an interruption between learning about a new concept and studying it in more depth makes it more likely to stick in your mind. This can be a more effective way to learn something than cramming.

Carey suggests taking a few short breaks to look at Facebook or similar: he suggests it refreshes your ability to focus and can allow your mind to keep processing problems 'offline'. At this point, he does back off a bit from the most rigid interpretation of his thesis, reminding the reader that you do at least have to have made some progress on understanding a concept before knocking off to play that videogame. This will give your brain a chance to do the offline processing.

Some of Carey's advice is less contrarian than he would like to think: he bases many of his suggestions in quite traditional research (although he often finds amusing ways to do this).

For instance, he points out that it is a good idea to do your learning in a variety of contexts, times of day and environments, as this helps you to access the memories of the things you learned in these various situations. He retells a story from the work of John Locke in which a man learns to dance in a particular room in which there is a trunk in a particular position. He finds after some time that he can dance well in that room, but elsewhere he can only dance if he has a trunk positioned in a similar location in the new room. And it has long been observed that this is true: some of the most ancient learning devices were memory 'palaces' in which someone mentally placed items to be remembered in different rooms of a building they knew well: again, the association of location and data allowed the mind to access the data more easily. (See Joshua Foer, p.137 for more on this.)

Some of the advice is entirely obvious: he cautions that you will learn better if you are motivated and a self-starter. Collaborative learning can help, especially if you discuss the material you are trying to learn. It's best not to be stressed in tests, partly because you will remember things better if you are in the same emotional state you were in when you learned them. It's also unsurprising to learn that getting a good night's sleep or napping can help your memory.

One of the more useful bits of advice is centred on the 'fluency illusion', in which we overestimate our ability to remember something in the future simply because we remember it now. The solutions to this include saying things aloud as you learn, and self-testing on a regular basis so that you can monitor how well you really are retaining the information.

Even if this isn't quite as contrary a book as it thinks it is, it is an enjoyable read and many people will find some useful practical nuggets that will help them learn, together with some useful explanations of the science that backs up that practical advice.

THE SPEED READ

Learning is easier if you vary the way you do it, and don't feel obliged to put in marathon sessions. Short bursts, with regular interruptions, can work better, as can roaming around learning in different places. Don't worry if you are forgetting things as it can strengthen your eventual memory, but do use self-testing to avoid falling into the fluency illusion and thinking you are doing better than you really are.

The Little Book of Thinking Big

Richard Newton, 2014

Publishers these days are increasingly fond of books that don't have too many words and include some pretty pictures or motifs. You might see this as evidence of a broad dumbing down across the industry . . . I couldn't possibly comment.

Anyhow, this is an example of the trend. It's under two hundred pages long, including full-page images for each chapter title, numerous doodles and a scattering of inspirational quotes on illustrated backgrounds. Flicking through it is somewhat like scrolling through your flaky friend's Facebook feed to read maxims and quotations they have reposted from elsewhere.

Having said that, it is quite a sweet little book in some respects (and it was a bestseller, so clearly some people love it). The chapters have titles like 'Have a Big Ego and a Small Ego', 'Float, Don't Swim', and 'Change Reality (. . . Don't Deny It'). The text is arranged in spaced, short paragraphs, often no more than a sentence: the effect is to heighten the expectation of profundity, whereas quite a lot of what is actually said is fairly banal (imagine *Jonathan Livingston Seagull*, rewritten as a smart-thinking book, and you've got the general idea). One typical paragraph runs, 'Because the bottom line is this: your imagination is now the limiting factor of your life. In the world of anything-is-possible, the outer limits of thinking big are the barriers of your life.'

The focus, of course, is on thinking big, particularly in business and with creativity. The author writes, for instance, 'That's what it took: ideas upon ideas, thinking big upon thinking big, to bring forth into the world his Big Idea, which would eventually not only make him a multi-millionaire but would also be one of those little noticed inventions that change the world.'

This is promoted as the way to encourage creative ideas and to improve your decision-making. Some of the most useful advice is about prioritising and choosing how to apply yourself. There are warnings against being an 'agreeing machine' rather than a 'deciding machine', because the more you allow others to dictate how you spend your life, the less control you are able to assert over it. This is an analogy the author has borrowed from Kurt Vonnegut, in the story 'Harrison Bergeron'. The plot is based on the idea that, in the future, people will be ruthlessly 'handicapped' to prevent the most intelligent, strong or attractive from having an unfair advantage. As a result, some people turn themselves into agreeing machines in order to be 'stupid on purpose'.

The metaphor of the 'Sargasso Sea of mind' is used to depict a becalmed state in which it is impossible for you to make progress. This happens when you give in to work pressures, gossip, the draw of social media, arguing with trolls and other mental drains. This all makes it hard to think big. You need to take a step back with respect to what you truly want to achieve, rather than focusing on the objectives that are foisted upon you by others.

Given that none of us is a tabula rasa, our life experiences make up a huge part of who we are. For this reason, you need to surround yourself with good, interesting people who will challenge your thinking and help to inspire you.

There is an interesting section on the value of idleness. The author carefully distinguishes this from laziness. Laziness is being unwilling to work hard or apply yourself. Idleness is regularly making the time and space to 'do nothing' and to allow yourself to reflect and take a step back. The universe is trying to show you things, to inspire you all the time: you can only be receptive to this if you give yourself time to notice what it is showing you.

There is also a neat distinction made between 'Yes, but . . . ' (which is really 'No') and 'Yes, and . . . ' which accepts a new state of affairs and starts the process of incorporating it into your life.

Inevitably, the book does descend into platitudes at times. By the time the last chapter is introduced with a page mostly blank, apart from the word '. . . . twang!' at the bottom, this can get a bit grating. But it would be unfair to dismiss Newton too readily. It's clear that if one were in the right mood, this could be quite an inspiring read, and one that could be consumed on a short train journey, to boot.

THE SPEED READ

The key to a better life is learning to think big. You need to clear your head, not get bogged down in the minutiae of everyday life, surround yourself with creative people who will challenge you, mash up ideas to create new ones, sometimes be idle (not lazy) to allow time to reflect, and most of all, allow the universe to show you things.

The News:
A User's Manual
Alain de Botton, 2014

Many of the more fatuous types of self-help books advise their readers to avoid the news because it will only make them feel stressed and distract them from ceaselessly imagining themselves as successful, beautiful and rich. That has always seemed like a fairly strange piece of advice, but it has to be admitted that in the modern world of wireless connectivity, social media and twenty-four-hour news, the influence of the news on our lives has become quite distorted and often negative. It is a pleasure, then, to read a far more leisurely, thoughtful account of the role that 'news' plays in our lives, and some calmer suggestions on how to deal with it.

Alain de Botton is a British philosopher and author who often applies philosophical ideas to everyday life. He is the co-founder of London's School of Life, which in itself is a bold experiment in teaching people to take a smarter, more philosophical approach to life.

The News: A User's Manual is a book-length meditation on what news is and how it affects us. De Botton begins by acknowledging the omnipresence of news in a wired world, and the way it makes us feel we have an obligation to treat the information that bombards us as crucial, urgent and unignorable. Think of the way that certain types of economic news are accompanied by the news anchor's frown, signalling to us what we should feel.

There is also the ever-present question of what constitutes 'news' in the first place. De Botton invites us to consider why particular stories were elevated to the front page, while others are downplayed. Why is 'Sydney man charged with cannibalism and incest' more appealing to readers than 'Tenants' rent arrears soar in pilot benefit scheme', for instance? He probes what this reveals – that readers simply are 'truly shallow and irresponsible' citizens, that the blame lies with journalistic convention, or if it is the news organisations' choice, based on the fact that they know that the spectacular, grisly, and weird will grab our attention more readily than mundane matters that directly affect our lives. He also addresses the question of how our stability and sanity enter the equation: 'If we were entirely sane, if madness did not have a serious grip on one side of us, other people's tragedies would hold a great deal less interest for us.'

The closest de Botton comes to defining what news actually is comes when he calls it the 'determined pursuit of the anomalous', which might be troubling if we needed a definition, but we all have a pretty clear sense of what news organisations regard as newsworthy and what they don't, and how shallow that conception is.

Instead, he focuses more on what the news could be in a better world. He talks of its current habit of 'randomly dipping readers into a brief moment in a lengthy narrative . . . while failing to provide any explanation of the wider context,' suggesting that the true news is more often like a novel, a long, unwinding story in which many small moments add up to a much wider meaning. For instance, a story about welfare cuts will be treated as though a sudden decision has had to be made about budgeting rather than acknowledging it is just another scene in the long story of how welfare affects lives, how we feel about helping those less fortunate than ourselves, and what role government should play.

Part of de Botton's self-proclaimed goal is to 'complicate' our idea of how we should respond to news. The idea that news should be more like novels comes with the implication that we should always bear in mind that we are being presented with a fragment of a story, a hundred-word extract from *Anna Karenina*, without the context needed to understand its true significance.

He is also good at capturing the range of conflicting emotions that can be triggered by certain types of news. A report on the stock markets might, for instance, impel us to admire the complexity of the business world, to wonder why we are supposed to care, to reflect on the absurdity or waste of our own working lives, and to wonder if we could be more successful if we changed our approach to life.

There is also a lovely passage reflecting on Flaubert's idea that wider access to information doesn't have only positive results, as now idiots can know things that only geniuses had understood in the past, without this making them any cleverer. As a result, 'The news had, for Flaubert, armed stupidity and given authority to fools.' This depressingly sounds like a description of every internet forum in the world.

One characteristic trait of de Botton is that he occasionally has interesting things to say about a subject but takes them on to bizarre conclusions. He gives a funny description of the pitfalls of celebrity culture and suggests a rather bland solution, in which we somehow make celebrities of genuinely admirable people, thus inculcating a desire in viewers to become better, more balanced, kind and productive. And in a passage about the problems faced by reporters trying to convey an accurate picture of life in a foreign region, de Botton ends up suggesting that the reporter might resort to creative writing, bending facts or changing dates on the basis that 'falsifications may occasionally need to be committed in the service of a goal higher still than accuracy'. This seems like terrible advice indeed.

Another flaw in the book is the omission of any deeper analysis of the political control and motivation of news organisations, as this is clearly one aspect that viewers should be aware of. But this is a gentle, enjoyable read nonetheless, and one that will challenge the reader to reconsider their attitude to the news.

---THE SPEED READ---

A meditation on what 'news' is, why and whether it matters, and the effect it has on our lives. Ranging from the question of why a grisly murder will take pride of place on the front page rather than something directly affecting our life, to the question of presentation, de Botton looks at whether there is a better way of doing 'news'. In particular, when most stories are the equivalent of a novel, what do we really learn from viewing sensationalised, bite-sized chunks of a far more complex narrative?

The Organized Mind:
Thinking Straight in the Age of
Information Overload

Daniel J. Levitin, 2014

This book begins with a touch of hubris, declaring on the first page, 'This is the story of how humans have coped with information and organisation from the beginning of civilisation,' which is a pretty bold claim. The more mundane truth is that it is a pretty interesting blend of business book, science or neurology book, and self-help book.

Levitin starts by discussing the way our brains and memories work, along with how we have evolved a wide variety of external and internal memory aids, from written notes to filing cards, memory palaces, and computers. His aim is to show how we can train ourselves to organise our brain and memory more efficiently.

This is a valuable aim in the modern world. We have never been more overwhelmed by data than now: from social media to twenty-four-hour news and from global culture and news on the internet to our hectic daily lives, we are living in the age of information flood. Levitin's thesis is that we can find ways to take back control of our lives.

He does this partly through some fairly practical suggestions: for instance, he points out that your brain has a daily limit for what it can process, which means that every time you look at a cute cat video on the internet, you have used up some power. He talks about the way that pressing 'Like' on social media is addictive, because it stimulates a hit of dopamine in the brain. He warns against multitasking, suggesting that trying to do several things at once is counterproductive, since you are just breaking your attention and consciousness into several pieces, making each less efficient. It is, scientifically and medically – due to the stress caused – better to concentrate on one task at a time. He is more positive about daydreaming, which he regards as one of the most productive activities your brain undertakes.

Levitin takes a more academic approach to explaining how the brain operates. For instance, imagine you have lost your keys. You would go through a process: first, you would double-check the pocket they are supposed to be in. Then you would try your other pockets on all your clothing, in decreasing order of how likely the pockets are to contain the keys. Then you would check other places: the front door, the car, the jar your partner keeps their keys in and so on. At a certain point, you would conclude that the keys are lost and move on to think about other things.

Each of these stages demonstrates one part of our attention system. The 'mind-wandering' mode comes with the discovery of the loss (along with a surge of cortisol and stress); the central executive mode kicks in when you look through your pockets; the attention filter works as you gradually widen your search; and the attention switch comes when you give up the search. The mind-wandering stage is as necessary as the central decision-making system, which is where the value of day-dreaming comes in.

In a similar way, Levitin talks about the ways in which you might get someone to think about a fire engine. You might give a verbal description of how it looks. You might ask them to name as many red things as possible. You might ask them to name emergency vehicles. Each of these involves using an attribute of the vehicle: it is red, it is an emergency vehicle, it has a certain appearance. The idea of "a fire engine" is represented in the brain by a cluster of neurons, as is each of these attributes: 'red things', 'emergency vehicles' and so on. The first cluster (representing the fire engine) is connected to the others, and is most strongly associated with the ones which are its most salient attributes. The easiest way to access the concept of a fire engine is to think about its most obvious attributes.

This is an example of the way that our memories attention systems work in practice. In daily life, we use a lot of external memory aids to assist it: lists, Post-it Notes, organisers and so on. Obviously, the more efficiently we use these, the more focused and organised our thinking will be. Levitin can get a bit nerdy about specific memory aids and has a tendency to wax lyrical about the effectiveness of 3-by-5-inch index cards, labels for drawers, pill dispensers, and nut-and-bolt displays from the hardware store.

His raft of low-tech solutions might seem out of place given their origins in the lab. But there is some really practical advice in here on everything from managing your workflow to organising your kitchen drawers. For instance, for the latter, he suggests having a junk drawer – remembering to sort it out occasionally.

He argues that you should value the long-term goal rather than the short-term rush, meaning you should remember to be organised and patient in the way you organise your internal and external memory. For instance, adopting a slow process of learning, in which you write things down on paper and say them aloud, in the old-fashioned way, will get better results than using Google and typing on your connected device, in the face of multiple online distractions.

Beyond the practical hacks and tricks, Levitin is keen to stress the medical and spiritual benefits of being organised. There is a kind of mindfulness that can be achieved only if you are not fretting about undone tasks and future problems, as you can then fully focus on the present.

Not everyone will agree with every suggestion Levitin makes, but this is an interesting and helpful guide to how the brain works and what is going on when you focus on a task, when you remember an important fact or when you allow your mind to wander.

THE SPEED READ

A bracing guide to focusing your attention and organising your information, both within the internal memory banks of your mind and through external aids such as notecards, labels, and storage solutions. If you want to achieve a Zen-like level of calm and to know where all your information is stored, then use this book to train your brain and to learn new habits that will help you to focus more clearly on the important tasks of the present day.

Sapiens:
A Brief History of Humankind

Yuval Noah Harari, 2014

Humans are only a small part of the story of life on Earth. The oceans were teeming with life from the Cambrian period (about 500 million years ago) onwards. The party on land started with plants about 450 million years ago, and a huge variety of amphibians and insects joined in shortly afterwards. The dinosaurs were the dominant species 250 million years ago, at which point the first tiny mammals were just starting to scurry around in the undergrowth, trying not to get trodden on.

Those mammals only evolved into the first primates fifty million years ago, monkeys thirty million years ago and apes developed around twenty million years ago. Humanity's ancestors only split from their chimpanzee and bonobo cousins seven million years ago and *Homo habilis*, the first humanlike creature, seems to have developed about 2.4 million years ago, whereas *Homo sapiens* dates back a few hundred thousand years. And this is where Yuval Noah Harari takes up the story, as this is an ambitious attempt to tell the entire story of our species in just a few hundred pages.

Our species has been through a series of revolutions: the cognitive revolution, seventy thousand years ago, in which we started to use our brains in much more inventive and creative ways; the agricultural revolution, about eleven thousand years ago; the scientific and industrial revolutions of the last few centuries; and the information revolution, which we are still going through. Harari is good at summarising these changes, and has some fascinating speculations about how the current biotechnological revolution might lead on to us becoming post-humans, immortal (or 'amortal') beings who need never die.

He is opinionated about these changes, but never less than challenging. He talks of the agricultural revolution, for instance, as 'history's biggest fraud', since it led to longer working hours, more starvation, more crowded and diseased living spaces, and more authoritarian societies. (He regards factory-farming and industrial agriculture, in particular, as evils of the modern age.)

Along the way Harari has much to say about how the development of language affected our ability to co-operate, communicate, and conceptualise abstract ideas, and the evolution of money, trade and empires. The really interesting thing about taking such a long view of history is that we can see how a mind that evolved to deal with being a hunter-gatherer adapted to the environment of high-rise living, with computers, debt, and social media. As Harari says, in spite of all the trappings of modern life, 'Our DNA still thinks we are in the savannah.' Harari depicts a race that is no happier than it was back on that savannah and, in many ways, worse off.

When he describes our possible amortal future, he is quick to speculate on the ways that an eternal life could lead to lassitude and boredom. He refers to the research showing how little correlation people's happiness has with their material wealth. Even lottery winners and millionaires can be unhappy, while the poorest slum dweller can find reasons to be happy in their daily existence. Harari's discussion of happiness can seem somewhat glib, but it is certainly interesting to ask whether or not 'progress' equates to 'happiness'.

Part of Harari's motivation is to challenge the version of history that paints humankind's story as one of gradual elevation from a brutish animal existence to the current dominant form of 'liberal humanism', which he describes as a kind of monotheism in which humanity worships itself.

There is, of course, a danger in taking such a broad-brush approach to history. He has a tendency to make sweeping statements and give opinions dressed up as facts: for instance, at one point he describes how modern institutions are reliant on belief in abstract entities, and states, 'Modern business-people and lawyers are, in fact, powerful sorcerers', which seems at best an overstatement. Experts in specific fields have pointed out oversimplifications and errors in some of his history – for instance, in the claim that the British defeat of the Ottoman fleet in the Battle of Navarino meant that Greece was 'finally free'. His claim that the nation-state is in inevitable decline in favour of a future global empire doesn't seem sufficiently grounded in evidence or facts.

Such quibbles are, however, inevitable in a work of 'big history' like this: what you get here is a slightly different view of humankind to that found in similar titles like Jared Diamond's *Guns, Germs and Steel* (see p.57). Each of these works in their own way challenges us to think differently about history. However it is approached, big history is fundamentally a challenge, as it asks us to ignore the small details of nations, monarchs, wars, and empires and to think about who we really are, in terms of how we evolved from a species of monkey into modern-day humans. Different authors will, of course, focus on different parts of the overall pattern, but taking that journey with an author as interesting as Harari is certainly a worthwhile endeavour.

---THE SPEED READ------------------------------------

A breakneck tour of the history of *Homo sapiens*, explaining how the various revolutions, breakthroughs and changes we have been through in the last seventy thousand years have shaped the modern world, along with speculation about the future of the species. Occasionally opinionated and not infallible, but a dizzying and challenging book that will make you see the people around you in a new light.

SmartCuts

Shane Snow, 2014

SmartCuts is a modern take on the concept of lateral thinking, first popularised by Edward de Bono (see p.30).

Shane Snow is a web entrepreneur and journalist. The aim of the book is to explore how people or companies achieve remarkable feats in surprisingly short amounts of time. He tells the stories of overnight successes, from startups that quickly reached billion-dollar turnover, to people such as YouTube tycoon Michelle Phan, and Jimmy Fallon, who quickly rose up through the showbiz ranks to host the *Tonight Show*.

Snow's take is that all of these people were behaving like computer hackers except, instead of taking shortcuts (that can lead to morally questionable quick gains) they used 'smartcuts', which get you to the target quickly but are sustainable. This involves breaking the usual rules and conventions and using lateral thinking.

The starting point is questioning assumptions about how a given problem should be solved. The conventional way of thinking may not be the best or most fruitful. Snow talks about the idea of 'hacking the ladder', pointing out that people who appear to be overnight successes have often already become strong or successful at something else. The road to success needn't be a linear one, and it can make sense to climb a certain distance up one ladder before swapping to a new one.

He also discusses the importance of mentoring, focusing on the example of Alexander the Great, who conquered a large part of the known world in a short space of time. He had one of the most esteemed mentors someone could hope for, Aristotle, as his tutor. Of course, such mentoring isn't always available or accessible; someone like Jay-Z didn't have anyone obvious, and substituted advice from extensive study of autobiographies and videos about his role models and how they attained their success.

The importance of getting good quality feedback on your progress is also stressed: Snow argues that people who are able to turn their egos off are often more successful as they are not just willing but keen to receive even negative feedback on their performance. They welcome it as it is the fastest way to grow, learn and effect change.

When it comes to momentum on your road to success, Snow emphasises the importance of small wins: if you have a long road ahead, minor successes along the way can help to keep up morale and enable you to keep up the momentum. It is important to set numerous small, attainable goals along the way. He also emphasises the importance of simplicity, arguing that creativity is encouraged by constraints. The smaller the scope of the problem, the more chance you have of solving it. 'There are a lot of great inventors and improvers in the world. But those who hack world-class success tend to be the ones who can focus relentlessly on a tiny number of things. In other words, to soar, we need to simplify.'

Some of the tropes are somewhat clichéd. For instance, Snow's focus on '10 × thinking', as commended by Google's Larry Page (instead of improving performance by 10 per cent, you should think how you can do it ten times better). From Google he also borrows the idea of '20 per cent time', although he points out this originated beyond the web. A company called Minnesota Mining and Manufacturing, aka 3M, let its workers spend 15 per cent of their time experimenting with new ideas (Google inevitably increased this). 3M's '15 per cent time brought us, among other things, Post-it notes.' Another slight weak point is that there is a somewhat Gladwell-esque focus (see Malcolm Gladwell review, p.93) on the elaborate justification of counter-intuitive ideas – such as, why kids shouldn't learn times tables.

On the whole though, this is a pretty inspiring book, which tells interesting stories about all kinds of things from the Cuban revolution, to SpaceX, and Skrillex.

---THE SPEED READ---

If you want to be seen as an overnight success or to turn a startup into a multi-million-dollar business you need to use 'smartcuts'. These are like shortcuts, but ethical and sustainable. The 'smart' bit comes from lateral thinking – don't accept your initial assumptions about a problem. Instead, simplify, streamline, and try to focus on excelling at solutions.

We Should All Be Feminists

Chimamanda Ngozi Adichie, 2014

One of the clichés of modern smart thinking is the ubiquitous TEDx talk: those brief lectures posted online on issues of philosophy, academia or general interest ('Ideas worth spreading', as their slogan promises).

You often encounter these as they are shared on social media. There is a tendency to the clever-clever in which a supposedly counter-intuitive but actually obvious statement is justified with some tricksy language to make the author look like a genius. But, to be fair, some of the talks are truly enlightening.

One such talk was given by the Nigerian novelist Chimamanda Ngozi Adichie in 2013. A simple but eloquent statement of the reasons why 'We should all be feminists', it has been viewed over five million times and was turned into this essay, in the form of a short book, in 2014.

It is a genuinely refreshing addition to the modern feminist canon, which all too often gets bogged down in complex, academic language and debates about intersectionality and the like. As a novelist and storyteller who grew up in Nigeria but also spent time in America, Adichie is more interested in the ways that small incidents and anecdotes can be mined for wider significance.

For instance, she tells about winning a competition as a child for which the reward on offer was being appointed 'class monitor', a role she coveted. Instead, one of the boys was given the role, even though the child who was appointed had no interest in performing the duty. The moral Adichie teases out from the story is that gender dictates how we should be rather than acknowledging us as we truly are.

It's worth pointing out that Adichie is extremely inclusive in her definition of feminism. She finishes the book by saying, 'My own definition of a feminist is a man or a woman who says, Yes, there's a problem with gender as it is today and we must fix it, we must do better. All of us, women and men, must do better.' And she is quick to point out that gender expectations affect men as well as women, arguing that as well as shrinking the horizons and ambitions of young women, they 'stifle the humanity' of young men. She describes masculinity as a 'small hard cage' in which we imprison men, and argues that men end up having fragile egos precisely because we expect them to be tough.

However, Adichie doesn't pretend that her views on feminism aren't partly rooted in anger: indeed, she embraces this. At one point she tells the story of a man telling her she shouldn't call herself a feminist or allow her anger about gender to show, because it will put men off from talking to her. Her response is that any man who would be deterred that easily is exactly the sort of man she doesn't want to know.

She is also extremely clear and direct in explaining why she uses the word 'feminism'. It is in response to those who argue that if you are arguing for equality of opportunity and rights between men and women, you should call yourself something like a 'humanist' or 'equality campaigner' rather than feminist. Adichie says that not using the word 'feminist' would be to pretend that it hasn't, across the centuries, been women who have been most excluded, most shackled and most disadvantaged by the disparity between the social roles that are allotted to men and women.

Adichie is powerful on the subject of shame, arguing that when we teach young women to cover themselves up, stay chaste, and not to express their desire, we are teaching them to remain silent about their true selves. Indeed, we are teaching them that pretence should be a normal part of their lives. And when, at the same time, we teach them to pander to the fragile egos of men, by being successful but not too successful, and by not intimidating or threatening men, we are teaching them to shrink their horizons, to lower their ambitions and to make themselves small.

The essay has become a widely distributed part of modern culture. Beyoncé included samples of Adichie in her song 'Flawless', every sixteen-year-old student in Sweden was given the book in 2015 and, in January 2017, one US bookshop gave it away free as a protest against Donald Trump's election as president.

This is an important book when it comes to the way we think about ourselves and about each other as it brings huge clarity to the issue of why gender equality is still an important issue that affects many everyday interactions. Adichie says, 'Gender as it functions today is a grave injustice. I am angry. We should all be angry. Anger has a long history of bringing about positive change.' But it is greatly to her credit that she has found such a positive, constructive way of expressing that anger.

THE SPEED READ

A short essay, based on her TEDx talk, in which the Nigerian novelist gives a concise, compelling statement of the ways in which gender expectations distort the lives and ways of thinking of both men and women, and argues optimistically for a better world in which a wider understanding of the virtues of feminism leads to a more balanced approach that allows both men and women to truly express themselves.

Black Box Thinking:
Why Some People Never Learn from Their Mistakes – But Some Do

Matthew Syed, 2015

Matthew Syed is a former table-tennis champion whose previous book, *Bounce*, was a fascinating look at the secrets of, and conditions for, success in sports and a variety of other fields. He takes a similar approach to that employed by Malcolm Gladwell, exploiting unexpected connections and surprising statements to take a fresh look at his subjects.

The basic thesis of this book is that the path to success often starts with multiple failures. You need to develop a more positive attitude to failure, learning from it and improving for your future. The title is derived from the device used to track all events in airplanes and which helps the industry learn from any failures. From the 1950s onwards flight recorders were gradually introduced, and now every aircraft has to have its own 'black box'. These are nearly indestructible and, after an accident, the data can be recovered so that the reasons for that accident can be understood, even after catastrophic crashes. While there are still notable exceptions, this is a major reason why the industry has a good safety record.

However, in other areas of life, people are less willing to face up to failure and learn its lessons. Basically, we don't like it when we fail, and we put less thought into understanding why we failed than we do to explain success. Syed's point is that this is the wrong way to go about things and can have terrible consequences in some key areas. He claims that the 'God complex' experienced by many senior doctors leads them to find it hard to admit they have failed, let alone invite their peers to review their performance and give them constructive feedback. We all need to embrace failure, to invite independent feedback, and use that feedback to improve our performance the next time around.

Syed uses the example of Juan Rivera to demonstrate this. He was wrongly convicted on a charge of rape in 1992 and spent thirteen years in prison. DNA testing had been available before the start of his sentence, but it was only in 2005 that the police used it to verify the genetic material that had been collected from the victim – this proved that Rivera was innocent. A lack of willingness to face up to the possibility that they had made a mistake thus led to his wrongful imprisonment lasting years longer than it needed to.

Syed discusses many of the cognitive biases and errors also discussed by writers such as Daniel Kahneman and the way that they lead us to overestimate our own abilities and to be overly certain about our beliefs. He suggests treating all conclusions and ideas as hypotheses that require continual testing against available evidence. As the saying goes, 'When the facts change, so does my opinion'. Failing to use this scientific method can lead to terrible outcomes – for example, bloodletting continued to be a common medical 'cure' over the centuries, in spite of the lack of any significant evidence that it actually worked. Instead of giving in to confirmation bias and assuming your methods or ideas are right, you actually need to go out and seek contrary or opposing evidence. It's only when that can't be found that you can decide you are in the right.

It's fine to make a mistake, so long as you learn from it. Syed references the inventor James Dyson, who made hundreds of prototypes of his bagless vacuum cleaner. His initial way of thinking was clearly flawed, but at each stage he patiently learned the lessons of his failures and worked his way on to final success.

Syed also talks about the difference between a fixed mindset and a growth mindset: the latter is the one that allows you to take responsibility for a mistake, own up to it and change, rather than giving in to the failure.

These ideas are based on many interviews and from other sources and are applied to a variety of areas including evolution, economic markets, and sports successes like David Beckham and the Mercedes Formula One team.

The main downside of the book is that it discusses a fairly narrow idea that is overstretched, and occasionally you feel that the point has already been made well enough. But Syed is an engaging writer who brings his concepts to life and the book is, on the whole, an inspiration, and one that has transformed many people's attitude to failure.

THE SPEED READ

Don't be downcast by your failures. Instead, embrace them: seek third-party analysis, learn the lessons and put those lessons to work in the future. The experience of the airline industry shows us that overall standards can be raised hugely by finding that 'black box', studying its contents and making adjustments to your approach in future.

Mindware:
Tools for Smart Thinking
Richard E. Nisbett, 2015

B efore he was known to a wider audience, the social psychologist Richard Nisbett already had a strong academic
following.

In 1977, he wrote what became a widely cited psychology
article with T. D. Wilson, 'Telling More Than We Can Know:
Verbal Reports on Mental Processes'. They discussed mental
processes when it comes to making choices and our emotions
can't be consciously accessed. Malcolm Gladwell (see p.93)
has described him as being one of his major influences.
Nisbett's 1993 book *Rules for Reasoning* is a terrific examination of experimental investigations of reasoning and, in
particular, the abstract rules we rely on, and whether or not
we can learn to change the unconscious choices we make. It is
to be hoped that some enterprising editor will be reissuing
this book soon.

Mindware is a comprehensive guide to the ways in which
our conscious thoughts don't always line up with the unconscious processes that are actually driving our decisions. The
result is that we don't always have access to the inner workings

of our mind, any more than we do to the inner workings of our kidneys or endocrine system.

Some trivial examples of this come from what are called 'incidental stimuli': people writing down items from a catalogue using a green pen will choose a higher proportion of green items than those using an orange pen; rhyming phrases are more persuasive than those that don't rhyme; we place more trust in a message if it is written in neat handwriting; the colour of the walls in a job interview room can affect the outcome. These are all proven experimental findings that reveal the complexity of our mental processes.

The cover blurb suggests that this book is a guide to how to think more clearly. That isn't strictly speaking accurate. Instead, it is an excellent guide to the many ways we get things wrong and always do, and how some awareness of these unconscious processes may help us to avoid unclear thinking in specific instances. In particular, Nisbett does talk about how to use rules of inference more clearly in everyday life, by learning how to frame a problem clearly.

Nisbett begins the book by talking about some of the unconscious processes that get in the way of framing everyday problems clearly. For instance, a lot of our thinking relies on 'schema': we associate a fancy restaurant with a range of attributes such as 'quiet' or 'elegant' and often reach snap decisions using only the schema, without actually looking into the details. The same applies when it comes to stereotypes – for instance, when we reach snap decisions about people we meet. The more we are aware of our reliance on schema and stereotypes, the more we can learn to put them aside when necessary.

It's also important to be aware of some of the irrelevant factors that affect people's decisions: judges are more likely to grant someone parole after lunch than in the morning. People who are given a warm coffee when they meet someone are more likely to see that person as affable and pleasant.

It's also worth knowing that the unconscious mind can often help us to resolve problems that we can't immediately solve consciously. People who sleep on a decision about a purchase tend to make better decisions and people studying for a test often perform better after putting the work to one side for a while before returning to it.

The second part of the book gives a fascinating exploration of cost/benefit analysis: Nisbett points out some of the mistakes we all tend to make when trying to weigh up pros and cons. For instance, we are prone to the 'sunk costs' fallacy, in which we 'throw good money after bad' rather than accept the time or effort we have already put in has been wasted. This explains why people will often continue to sit through a terrible movie rather than walk out of the cinema and, at a more serious level, expensive government projects can be continued when the rational decision would be to walk away. Nisbett also looks at our regular failure to factor 'opportunity cost' (the notional costs incurred by turning down other opportunities) into the equation and the ways in which 'loss aversion' (the way in which our fear of a certain loss can outweigh the hope of an equal but opposite gain) can lead us to poor decisions.

The book also goes into great detail about statistical errors and the sorts of irrational fallacies that afflict people who design and carry out experiments. There is a particularly interesting discussion of 'multiple regression analysis', a technique often relied on in the *Freakonomics* books (see p.97), which Nisbett refers to as 'eekonomics' with his tongue in his cheek. The basic idea of multiple regression analysis is to 'correct' the data, removing unwanted or irrelevant variables, so that the resulting data set only varies with the dominant variable we are trying to isolate. It is a notoriously unreliable process that can often amount to little more than guesswork. It is hard to truly identify and eliminate every factor that might affect a study.

Nisbett's style can be dense and at times his writing is quite technical and this book is not always as accessible and easy to read as some more populist titles. If you have read Daniel Kahneman's *Thinking, Fast and Slow* (see p.144), you might feel that this is covering much of the same ground. But it is undoubtedly an interesting read, and an authoritative book by an expert in his field.

THE SPEED READ

We don't have conscious access to all the unconscious processes in our mind, which are often the triggers for our decisions and choices. This is a comprehensive guide to the many ways that we make bad decisions, misunderstand situations and get the data wrong, but also a reminder that our unconscious mind can and does work for us, if we can let it.

Wired to Create

Scott Barry Kaufman and Carolyn Gregoire, 2015

The psychologist Scott Barry and the journalist Carolyn Gregoire wrote a *Huffington Post* article called 'Eighteen Things That Creative People Do Differently'. This led to this book, in which the two authors investigate ten 'habits of mind' that are features of creative people, including attributes such as daydreaming, and having an intuitive sensitivity.

They start with a look at historic and recent research on the subject of creativity and their general thesis is complex and nuanced. They argue that creative people often have contradictory, messy, and paradoxical features in the way they work. These can't be reduced to a simplistic set of rules. In spite of this, they argue, you can foster your own creativity if you embrace certain aspects of life, such as solitude, spontaneity, self-knowledge, and an understanding of tragedy.

Their overview of the research leads them to dismiss some traditional ideas such as the left/right brain theory (see p.104 [Daniel Pink]), and a correlation between creativity and IQ or divergent thinking (meaning spontaneous, non-linear thought processes). In fact, the most closely related thing to creativity is being open to new experiences and new ideas: 'It's the thrill of the knowledge chase that most excites them.'

The tricky thing is that this initial thrill is often followed by the grindingly slow execution of an idea. 'Creativity is a process that reflects our fundamentally chaotic and multi-faceted nature. It is both deliberate and uncontrollable, mindful and mindless, work and play.'

Each chapter focuses on one of the ten habits of mind. The first is the way that people with 'messy minds' are open to contradictory ideas. This is one of the things that allow creative people to break or remake the rules of their activity. The authors discuss the amount of energy that can be wasted along the way – for instance, Picasso made a huge number of sketches for *Guernica*, revising elements that eventually appeared unchanged in the final version, while going down dead ends with other experiments that he abandoned.

The second habit is, unsurprisingly, that creative people can be deeply passionate both in their own lives and about their work. The psychologist Martha J. Morelock has described the passions of creative individuals in terms that are reminiscent of the language of addiction, suggesting their focus derives from a real need to engage with their subject.

Without getting into too much detail, the other eight habits are:

- A sensitive and perceptive nature (that might contrast with an apparently arrogant and self-confident persona)

- Openness to new experiences (also known as 'psychological plasticity', stimulated by the production of dopamine, which is also associated with vivid dreaming, a common experience for creative people)

- A willingness to give the mind over to daydreaming and intuitive thinking (Carl Jung dealt with his emotional issues by allowing his mind to wander in a conversation between his subconscious and conscious mind)

- A liking for the experience of solitude on a regular basis (as embraced by thinkers as disparate as Immanuel Kant, the Buddhist monk Matthieu Ricard, William Wordsworth and Virginia Woolf)

- The ability to turn bad experiences into opportunities (Paul Klee redoubled his artistic endeavours after being diagnosed with a terminal disease)

- The ability to experience and practise some form of mindfulness or meditation (as shown by creative thinkers such as Ruby Wax and Steve Jobs)

- A willingness to divert from established routines and try something new

- The ability to try something new and not care if it doesn't succeed (which connects to the idea expressed in *Mindset* (see p.152) that the ability to see failure as an opportunity is a key to growing as an individual)

This isn't a book that gives easy answers or formulas, but it is a fascinating exploration of what is really going on with creative individuals, and how you can foster your own creativity.

THE SPEED READ

Creativity is associated with a wide variety of traits that can often seem paradoxical and contradictory. Creative individuals often have messy minds and take circuitous paths to their breakthrough ideas, but they also have the ability to see an idea through. Openness to new experiences correlates strongly to creativity and in this book are nine more habits of thought that can help you develop your own creativity.

The Descent of Man

Grayson Perry, 2016

The English artist Grayson Perry is best known for his cross-dressing and the fantastical pottery he makes, which might not make him the most obvious candidate for writing a book about the problems of toxic masculinity. He is, however, quick to point out that he has grown up with some traditionally masculine traits, such as a love of James Bond and motorbikes, and has a hugely competitive streak.

He starts this book with an anecdote about a young boy struggling to work the gears needed to get his mountain bike up a hilly path. The boy is clearly upset and shouting for his dad, who is standing a few hundred yards away with arms folded and a furious look on his face. It's the same face that many men will recognise as the look of disappointment from a certain type of macho, competitive dad.

Cleverly inverting the title of Charles Darwin's famous work, *The Descent of Man* is a brief investigation of the whole idea of masculinity. Perry points out, 'All over the globe, crimes are committed, wars are started, women are being held back and economies are disastrously distorted by men, because of their outdated version of masculinity.' At the same time he discusses the idea that men tend to be distant and emotionally withdrawn and the ways that 'default man' still tends to aspire to traditional masculine tropes.

The book isn't hugely original: for instance, pointing out that our stereotypical image of a world leader, professor or judge tends to be a man is a pretty obvious idea. Nor is it hugely insightful to suggest that men need to work less on their muscles and more on their intuition. However, Perry has an engaging and enjoyable writing style, and often creates hugely memorable images. For instance, he describes ties as 'colourful textile phalluses' around the necks of establishment males. And he suggests that grey suits emphasise neutrality, whereas it might give a more accurate image of the beliefs of someone like then-politician George Osborne if we pictured him dressed as a combination of Flashman and the Grim Reaper.

Perry is interesting on the subject of clothes in general. For instance, he talks about the ways that male fashion often includes as much frippery as female fashion, but with men it is often disguised by the functional side of such items as zips and buckles.

He essentially suggests that we need to rethink the definition of masculinity in the current world. Noting the popularity of programmes about survival in the wild, he suggests that the real survival skills we need now are those that help us to find a good home or school in a capital city, or achieving the kind of emotional intelligence that Barack Obama embodied.

However, he is quite understanding of the problems faced by men, even by the extremist tendency who hang out on forums talking about men's rights and red pills (representing the unvarnished truth of reality, as posited in *The Matrix*). He points out that they have inherited a narrative that emphasises male dominance, but mostly live out their lives in a state of frustration and servility. He also reaches some rather surprising conclusions, suggesting that we reintroduce national service as a way of allowing young men to come to terms with their masculine energy.

One frustration is the slightly confused approach Perry takes to the construction of masculinity. If we assume that social factors shape our ideas of masculinity, then smart thinking and self-examination can be seen as ways to amend and revise our own masculinity. If, however, natural and evolutionary factors are more important, then it is perhaps a more intractable problem. Perry frequently implies that social factors are key and that men can change, but at other times he seems to assume that there is such a thing as 'typically masculine' behaviour.

The book also becomes a bit repetitive in the latter stages. However, it is a good read by a fascinating man who is at his best when he is talking about his own life and experiences (and some of his artwork is included in the book).

He concludes with a list of rights that men should aspire to, which sums up the most important parts of his argument nicely. He suggests they should aspire to the right to be vulnerable, weak, to admit to being wrong, to be intuitive, to be uncertain, to be flexible. And most importantly he suggests they should have the right to not feel shame at possessing or acquiring any of these attributes.

┌─ **THE SPEED READ** ────────────────────────────────┐

A cross-dressing potter makes the case for using smart thinking and self-examination to rethink what it means to have a Y chromosome and to be a man. He concludes that, if we could all calm down a bit and be more intuitive, less overconfident and less angry, the world might be a better place all round.

└───┘

Get Smart!
How to Think and Act Like the Most Successful and Highest-Paid People in Every Field

Brian Tracy, 2016

Brian Tracy is the author of more than seventy books, including successful titles such as *Earn What You're Really Worth*, *Eat That Frog!*, and *The Psychology of Achievement*. He is also CEO of Brian Tracy International, a San Diego-based company selling counselling on selling, leadership, self-esteem, success psychology and much more. Whether that makes you see him as a motivational guru or a snake-oil salesman is probably down to your cognitive biases.

The heart sinks further when the blurb for the book rolls out the old chestnut we 'only use 2 per cent of our mental ability' and promises to use the latest brain research to allow us to unlock the rest. This is not only hackneyed, it is also scientifically askew.

More seriously, there is some perfectly acceptable advice in this book. Tracy exhorts the reader to work out what their goal is and how to get there. He suggests you copy the habits of role models, and try to change your habits one at a time.

Work hard and don't think forty hours a week or five days a week is enough. The book bigs up the power of positive thinking and advises you to get rid of negative mental baggage. It tells you that the sooner you invest your money, the more the interest will add up.

It's hard to argue with most of this, though most people will balk at the workload and some will find the focus on positive thinking to be tiresome. The main problem is that there is very little in this book that is fresh or that differs from hundreds of other peppy books on smart thinking. There is also a huge amount that the book shares with the most tedious of positive-thinking titles.

For instance, consider the following habits that Tracy attributes to successful people: they look at things in their entirety and in the long-term. They make time and space to think. They continually learn new stuff and aim towards their goals (which they will, of course, have written down). They have a positive, flexible and creative approach and they focus on results. All pretty obvious and, while Tracy expands on each habit and gives some occasionally interesting hints on how to achieve them, it's not really groundbreaking stuff.

Let's give him the benefit of the doubt and outline a couple of the brighter moments. He writes quite well on the problem of mechanical thinking (being stuck in set patterns of thought that don't help you move in the right direction). To combat this, he suggests taking certain precautions: be clear in setting goals, but flexible in how you achieve them. Stay focused, don't break your time down too much, and do a small number of things well rather than many badly. And design your environment to minimise distractions. For instance, turn off your email so you can concentrate on the task at hand.

All reasonable enough advice. Elsewhere he makes a neat comparison between the human mind, constantly fizzing with thoughts, and the bubbles in a champagne glass. The fizzing can produce excitement and fun, but the bubbles pop all too soon, leaving behind a flat drink. This connects to advice about making time each day to focus on your goals while the bubbles are still fizzing.

Anyhow, you don't get to write seventy books without knowing how to turn out a few nice ideas and analogies. But there are also passages like this which are weak to the point of meaninglessness: 'Success is the ability to solve problems as well. A goal or an objective unachieved, in any area, is merely a problem unsolved. This is why a systematic approach to problem-solving, one that works at a higher level and more consistently, is absolutely vital for you to achieve the maximum success that is possible for you.' Right . . .

There are occasional bright moments, but on the whole this is a book that feels like it could have been written on autopilot, or by a random, positive-thinking, sentence generator. Best to skip it and read some of the many better books on the same subject.

---THE SPEED READ---

In order to be successful, you need to think like a successful person. Find out what mental and practical habits your role models have and emulate them. Um . . . that's about it.

Grit

Angela Duckworth, 2016

Do you want to know the secret of success in publishing . . .? OK, I don't really know the answer to that one. But I do know that giving a TED talk is a great shortcut to getting a book deal. Especially if the talk gets over five million views on YouTube.

The problem is that what makes for a great talk doesn't always make for a great book. This is a case in point. Duckworth is a serious academic who has done some interesting work, but this feels like it has been geared to find as big an audience as possible. This tends to mean it has to sell magic beans: in this case, the beans are the wonder ingredient of 'grit' being a sure-fire path to success.

'Grit' is defined here as a combination of passion and perseverance. We often see talent as the prerequisite for success: Duckworth argues that talent alone isn't enough, because there are no shortcuts to excellence. 'Developing real expertise, figuring out really hard problems, it all takes time – longer than most people imagine . . . you've got to apply those skills and produce goods or services that are valuable to people . . . grit is about working on something you care about so much that you're willing to stay loyal to it . . . it's doing what you love, but not just falling in love – staying in love.'

This is a fair point, but in spite of the amount of – mildly – interesting research presented, along with accounts of meetings with successful people from various walks of life, it becomes pretty repetitive.

The most important points of the book can be summarised fairly briefly. You need to discover what you have a passion for. Then you need to put in a lot of very conscious practice because, 'as much as talent counts, effort counts twice'. Of course, this all works because we are human, and as such we need a purpose in life. And we need hope, the ability to be able to picture the light at the end of the tunnel, in order to be able to do that gritty thing and stay the course all the way to the finishing line. (Apologies for the mixed metaphor: I've been reading some cheesy books lately . . .)

There are a few other good points made in the book. Duckworth writes that even elaborate tasks can be made more doable by breaking them down into smaller ones. The difficulty comes when we lose sight of what each individual step is helping us to achieve and are unable to focus as a result.

There is also a revealing look at the difference between optimists and pessimists. Everyone encounters a similar amount of adversity in their lives. The difference is that optimists give explanations of these setbacks that focus on temporary, specific causes. By contrast, pessimists are prone to assuming the causes of their adversity are all-pervading, nebulous or permanent.

It is also reassuring that Duckworth points out that grit isn't an innate quality, but something we develop through our lives: 'Grit grows as we figure out our life philosophy, learn to dust ourselves off after rejection and disappointment, and learn to tell the difference between low-level goals that should be abandoned quickly and higher-level goals that demand more tenacity. The maturation story is that we develop the capacity for long-term passion and perseverance as we get older.'

And finally, it is worth bearing in mind the fact that you don't always see the grit that other people are showing. It may be obvious when you see an athlete putting in many hours of training: but the grit it takes to finish a PhD thesis, or to become a successful magician is mostly applied behind closed doors. 'Nobody wants to show you the hours and hours of becoming. They'd rather show the highlight of what they've become.' Any time you are prone to thinking someone has achieved great things through luck or innate talent, remember you may not be witnessing the hundreds of hours of hard work that created that luck and talent.

Presenting grit as the magic ingredient is something done by many 'How to be successful' manuals. Duckworth tells a story of a waitress who learns every job in the restaurant she works in and is rewarded when she is made manager. And, of course, that could happen, but so could many other scenarios in which the hard work isn't rewarded so handsomely. The author also gives a flattering depiction of Jamie Dimon, the financier, which suffers from survivorship bias (in this case, the failure to remember that the survivors in a given situation may simply have been the lucky ones rather than geniuses) and a tendency to believe people's accounts of their own successful path. As many readers have noted, if you focus too much on grit as a panacea, it can also lead to you failing to see how much inequality in outcome is caused by inequality in opportunity and to end up believing that society is far more meritocratic than it really is.

Grit certainly is an important attribute, and in many situations having that combination of passion and perseverance can indeed be a positive thing, but you don't really need to read 464 pages or so in order to understand that message. You might be better served by deciding what it is you really have a passion for and getting on with doing something about it instead.

THE SPEED READ

Grit is a combination of perseverance and passion: talent is useful, but without grit you will have less chance of success. Find out what you have a passion for, put in the long hours needed to master a skill or role and keep showing grit. Have you written three hundred pages of a book? Not enough! Show some grit – and keep writing!

Messy:
The Power of Disorder to Transform Our Lives
Tim Harford, 2016

If you've ever been oppressed by someone waving Marie Kondo's *The Life-Changing Magic of Tidying* at you, demanding you KonMarie your living area (say 'thanks' to your many possessions and chuck them out), you may be drawn to this book purely from a sense of contrariness. What could be a better antidote than a book proving that disorder and messiness are actually vital elements of creativity and success?

And to some degree that is exactly what this book is for: Harford writes, for instance, 'Often we are so seduced by the blandishments of tidiness that we fail to appreciate the virtues of the messy – the untidy, unquantified, uncoordinated, improvised, imperfect, incoherent, crude, cluttered, random, ambiguous, vague, difficult, diverse or even dirty.'

But from the start it is worth including a caveat: the title of this book is actually slightly misleading. As is so often the case with books in this genre, it feels like a publisher has been won over by the flashy concept of a book on a slightly unorthodox subject, only for the author to present a more nuanced, complex argument when they come to write the actual work. In this case, Harford (best known as the *Financial Times'* Undercover Economist) is often writing more about control, spontaneity and autonomy. There is a telling anecdote to this effect early in the book.

In 1993, Jay Chiat, head of the Chiat/Day advertising agency, commissioned the architects Frank Gehry and Gaetano Pesce to design new 'playful, zany and stylish' offices for their Los Angeles and New York branches. The result was the removal of cubicles, offices, and traditional desks, in favour of elements such as a four-storey pair of binoculars, murals of red lips and chairs with springs instead of feet.

When Frank Duffy, another architect, saw Pesce's creation he commented, 'Perhaps its gravest weakness is that it is a place where "play" is enforced on everyone, all the time.' The point is that such a space does not suit everyone, and making fun and chaos compulsory is no more effective than banning them.

Similarly, Harford talks about Building 20 at MIT, a notorious source of extraordinary scientific ideas, home, for instance, to the invention of computer hacking and the first videogame. Scientists in this nondescript, functional building were subject to very little supervision and as a result it could be chaotic. When experimenters wanted to build an atomic clock, they simply knocked through two floors to accommodate the equipment, leaving pipework, cabling and broken walls on display. Again, the real point of this story is not the mess, but that the scientists had control over their own space.

Having said that, Harford does use some more obvious examples of messiness leading to success. He talks about the way Jeff Bezos ploughed into the launch of Amazon and was from the start in a state of chaos, in turn leading to creative solutions to the problems he had created. He also points to the

German field marshall Erwin Rommel, who was known for his chaotic planning and willingness to disobey orders on a whim. As a result, the Allies found him a truly difficult enemy: they had broken the German codes, but this still didn't allow them to guess what Rommel was going to do next.

Harford also points to Brian Eno's use of his 'oblique strategies' – unusual, disruptive methods used to affect the process of the musicians he was producing, such as telling them to play each other's instruments – and the remarkable results this could produce. Paul Erdos was a brilliant, nomadic mathematician, and Benjamin Franklin was a famously messy inventor and writer (Harford notes that Franklin regretted not being more organised late in life, feeling he could have achieved more, but takes the opposing view, that Franklin's disorderliness was clearly related to his impulsively creative mind). Harford also spends some time looking at the virtues of allowing collaborative endeavours to be more chaotic and spontaneous.

It has been widely noted that Harford's book relies too heavily on male role models, and in particular on rich and successful people who can expect tolerance from those around them for being late. Ignoring appointments (which he also praises at one point) might work for a millionaire, while a low-wage worker could be sacked. A low point of this tendency comes in Harford's praise for Donald Trump's 2016 presidential election tactics.

He compares Trump's penchant for disorientation to other figures discussed in the book such as Bezos and Rommel, as well as his chaotic use of social media. He does go out of his way to point out the virtues of immigration, in an attempt at providing balance. But there is something a bit jarring about the way he talks about Trump's social media activity, noting fairly blandly the 'self-selection' that means that the creators of 'blue tweets' and 'red tweets' (in other words Democrat and Republican) that follow in the wake of an event like Michael Brown's shooting by the Ferguson police in 2014 rarely interact with each other. Rather than noting the dangers that arise from such a divided nation, Harford's take is that the problem is that users of social media choose 'a tidy corner', one they feel comfortable in. This seems a bit glib, and is an example of the way that pursuing a single theme through a book can lead to odd conclusions.

This is an enjoyable book, and as a hymn to the virtues of autonomy and spontaneity, it has great value, but in the end you simply can't reduce all problems to the conflict between messiness and tidiness, which is the source of the book's greatest flaws.

THE SPEED READ

Messy, disruptive, spontaneous men (yes, mainly men here) like Jeff Bezos, Brian Eno, Michael Crichton, Rommel and, er . . . Donald Trump, can baffle their opponents and make creative leaps that would evade someone with a tidier mind. We need to overcome our neurotic obsession with keeping things tidy and orderly and allow a bit more chaos into our lives.

Peak

Anders Ericsson and Robert Pool, 2016

Anders Ericsson is the Swedish psychologist Joshua Foer consulted when he was learning to become a champion mnemonist and his ideas are explored in Foer's *Moonwalking With Einstein* (see p.137). This book, written in collaboration with science writer Robert Pool, is a more detailed exposition of his thoughts, essentially a guide to improving your performance in more or less any kind of activity.

Ericsson has interviewed and tested a wide range of people who are known to be gifted or champions within their field, coming to understand the traits they have in common. He starts with a meditation on what it means to be gifted. As a child Mozart was famous for having 'perfect pitch', which meant that he could correctly identify a note, whatever instrument it was played on. For instance, if he heard a violin playing a single note he could name it as, say, the A flat two octaves above middle C. Perfect pitch is extremely rare: less than one in 10,000 people has it. For a long time, it was thought that it was a 'gift' that some people just had.

However, over the centuries, our understanding of perfect pitch has changed. It was noted that it was extremely rare for anyone to have it unless they had some musical training in early childhood. Then it was found to be more common in Asian individuals, but only those who spoke a tonal language (in which the pitch of a word can change that word's meaning). Finally, an experiment at Ichionkai music school in Tokyo by the psychologist Ayako Sakakibara demystified the subject.

He put twenty-four children between the ages of two and six through a training course in which they had to learn to identify chords played on a piano. The daily training sessions were short and continued until the children could correctly identify the set of chords he had selected. This was quickly accomplished by some children, while others took more than a year. But, in spite of this variation in aptitude, it was found at the end of the study that every child had acquired perfect pitch. The attribute is not a gift, but a skill that can be acquired, given the right training.

Ericsson's message is extremely simple: to improve your skills in any activity, you need to do the right sort of exercise, carried out for a sufficient quantity of time. This is what makes the difference between an amateur and an expert. It is not to deny that genes, opportunity, motivation, and other natural attributes contribute to the skills people acquire. But no matter how blessed you are at the outset, you will never become truly expert without this quantity and, more importantly, quality of practice.

Ericsson distinguishes three types of practice: naïve, which is not planned or thought through; purposeful, in which you set up a series of planned steps, keep pushing yourself beyond your comfort zone, use mental representations of what you want to achieve, and stay motivated; and deliberate, which is like purposeful practice but also (at least initially) involves the guidance of a teacher. It isn't always easy to find the right person to teach. You need to search for someone who has experience with people like you, you need to be prepared to change your teacher or mentor over time, as needed . . . and you may even need a bit of luck, as the right teacher isn't always accessible. You can instead study expert performers in your field to find out how they got so good. Whether it be the champions of a sport or your personal heroes, you should learn as much about their journey from beginnings to expert level as you can. Then you can incorporate their routines, targets and systems.

Ericsson warns against the danger of the plateau. It's easy to get to a level where you feel you are performing as well as you can, that you are in the 'flow' and there is no scope for improvement. To break away, you need to push yourself. 'This is a fundamental truth about any sort of practice: if you never push yourself beyond your comfort zone, you will never improve.' Foer tells of when he attained the skill of memorising cards at a certain rate (measured by a metronome). Ericsson's advice was that he should deliberately set the metronome at a faster rate – by 10 to 20 per cent – and then try to keep up until he failed, and repeat the exercise: the result was a rapid improvement in performance.

Mental representations can also help: experts in a field spend time mentally viewing all the options, and the steps they will need to take. As a result, their neural networks adapt and change, and the processes needed for the activity become more and more natural.

Ericsson is a true expert and the fact that he has met with so many top-level performers and compared their practices and mental attitudes gives this book real gravitas. There are books that promise magic bullets and secret formulas, and by contrast it may be boring to say that it is all about hard graft, in the right way, for long enough . . . but at least it's the honest truth.

┌─**THE SPEED READ**───────────────────────┐

Improving your performance is not just about 'grit', not just about doing ten thousand hours of practice, and certainly not just about positive thinking. The true champions and experts have spent long enough practising, in the right way, with the right guidance. And they have constantly challenged themselves to get out of their comfort zone so they don't get stuck at a certain plateau.

└──┘

Weapons of Math Destruction:
How Big Data Increases Inequality and Threatens Democracy

Cathy O'Neil, 2016

This is an excellent antidote to the hubris that is occasionally displayed in *Big Data* by Viktor Mayer-Schönberger and Kenneth Cukier (see p.165). Cathy O'Neil is the mathematician behind the Mathbabe blog and has also spent some time working at the hedge fund D. E. Shaw, which gave her a front seat view of the 2008 financial crisis. This left her with an understandably negative view of the financial sector and, in particular, the role that mathematical models played in that crisis.

Of course, there is a degree to which the use of models was as a smokescreen for dubious practice, as it was for instance when non-performing loans were bundled up into instruments that could be awarded a high credit rating and sold on. But a large part of the problem did come from the way that quantitive analysts (quants) relied on mathematical algorithms. These gave them an exaggerated sense of confidence that they were not taking any risks when, in fact, the risks were tremendous.

That story is, of course, well known, but the interesting thing is the similarities that O'Neil observed when she began working in other industries such as risk management and later as a data analyst at an internet media startup. 'I saw the same pattern emerging that I'd witnessed in finance: a false sense of security was leading to widespread use of imperfect models, self-serving definitions of success, and growing feedback loops. Those who objected were regarded as nostalgic Luddites.'

O'Neil brings a real sense of moral outrage to her descriptions of some of the ways that the models are affecting the lives of ordinary people, such as employees with zero-hour contracts who constantly have their part-time schedules juggled around to maximise profits. There are also the customers of credit card companies who had their credit limits arbitrarily lowered after the companies used criteria such as 'People who shop at Walmart' to assess risk.

She goes on to consider a number of other ways in which the models affect our lives. For instance, she explores the way that rich families and colleges game the system to trick the most common algorithm into favouring them for entry. This has a profound impact on who gets into the elite colleges of the USA.

Scammers, payday lenders and for-profit educational institutions are among those who exploit targeted advertising to reach the most vulnerable in society. It's far more worrying than you might imagine if you only saw, say, your favourite shoe store sending you details of their new lines.

These algorithms can disadvantage entire communities: predictive policing provides a way for police departments to cut costs, rationing the amount of law enforcement deployed in certain areas; other algorithms automatically sort job applications and effectively discriminate against residents of certain areas or racial groups; and access to mortgages and insurance is hugely affected by the algorithms that rate the risk attached to certain areas or demographics. Algorithms can rate excellent teachers as 'poor' based on some faulty criteria for 'success'.

Incidentally, O'Neil is fascinating on the subject of how the algorithms are sometimes just reproducing human prejudices. She writes, 'Racism, at the individual level, can be seen as a predictive model whirring away in billions of human minds around the world. It is built from faulty, incomplete, or generalised data . . . needless to say, racists don't spend a lot of time hunting down reliable data to train their twisted models.'

The chapter on Facebook has been somewhat overtaken (albeit vindicated) by events in the last few years, when their applications for user data have been subject to increased scrutiny. But the book makes it clear how dangerous it is that we give so much of our personal data to the big corporations, whether it be Facebook, Apple, Google or whoever. The dystopia O'Neil fears is one in which inequality rises as a lucky minority are enabled to cash in on the data economy and make fortunes (which they believe they are entitled to) while those deemed 'losers' are tagged that way for ever.

In terms of smart thinking, there are a few morals one can take from the book. One is to learn what algorithms affect you and to try to discover how they are being used and whether they are disadvantaging you. Another is the fact that, no matter how immoral their possible uses, algorithms are here to stay, so it is important to find ways to build models that incorporate as much morality as possible. In the end, most readers will come away determined to give as little data as they can to the corporations who increasingly dominate our lives.

THE SPEED READ

Big data is taking over the world and this is a scary thing. Corporation drones overestimate the power of the models they use, even while they don't truly understand them, and the result is inequality, unfairness and stereotyping. This has malign effects on everyone from college applicants to teachers, and from credit applicants to the victims of crime.

How to Think:
A Guide for the Perplexed
Alan Jacobs, 2017

Why are so many people so bad at thinking? This is the essential question asked by an English professor and cultural critic in this engrossing little book.

The short answer is that we grossly overestimate how good we are at it. 'All of us at various times in our lives believe true things for poor reasons, and false things for good reasons.' Jacobs points out that we are rarely as open to new ideas and to having our assumptions challenged as we believe we are. And all our ideas are rooted in what we see in our 'community', whether that be a friend, group or the wider community of people whose beliefs we share. 'Thinking independently, solitarily, "for ourselves", is not an option.'

Part of Jacobs' mission is to challenge the kinds of tribalism that have led to the era of 'fake news'. The idea that we are always thinking 'together' rather than in isolation, means we also need to think carefully about how we deal with opposing ideas. He points out how much of the division in society, whether it is between Brexit's remainers and leavers in the UK, Trump supporters and haters in the US, or orthodox Christians and academic critics, is a result of people intentionally mis-representing the views of their 'opponents'. Only by really hearing what the other person is saying and by being open to the possibility we might be wrong can we make progress towards the truth (not to mention towards having a less shouty internet).

We need to start actually listening. And if we think we understand the other person's point of view, we need to think about whether or not we are understanding it in a way they would recognise. People with ideas different to us aren't necessarily evil, and it can be a positive thing to surround ourselves with people who do think differently, particularly if they are good thinkers themselves.

Jacobs echoes Daniel Kahneman's terms (from *Thinking, Fast and Slow*, see p.144) for the two thinking systems: System 1 is intuitive thinking – immediate and fast – while System 2 is a slower kind of conscious reflection. Jacobs quotes the psychologist Jonathan Haidt making the comparison of System 1 to an elephant and System 2 to a rider. We mostly career through life using the elephant, an extremely powerful creature. But the rider can more finely control the elephant.

'Intuitive thinking is immensely powerful and has a mind of its own, but can be gently steered [by a rider] especially if they are skilled and understand the elephant's natural tendencies,' writes Jacobs. He suggests we understand how the elephant functions in order to become better riders. We need to be able to deploy System 2, especially when there is an anomaly or something discomforting in our intuitive responses.

The root of all of this is that we don't really want to think clearly. It can make us feel uncomfortable and complicates our relationships with people. And to do it well, we need to do it slowly which is increasingly difficult when we are bombarded with social media and twenty-four-hour news. System 2 thinking is not quick, but deploying it to its best effect can help us to avoid the kind of partisan squabbling and cognitive errors of those who are quick to communicate without thinking hard first.

Jacobs is a genial, funny author, describing, for instance, the many cognitive biases we suffer from, including anchoring, confirmation bias, and the Dunning-Kruger effect as 'the infinitely varied paths we can take to the seemingly inevitable dead end of Getting It Wrong'. Jacobs is also amusing on the dangers of being so obsessed with having an 'open mind' that it never closes on any actual conclusions. Using a broad range of references, ranging from novelist Marilynne Robinson, philosopher John Stuart Mill, and writer C. S. Lewis, to basketball star Wilt Chamberlain, this is a bracing, concise summary of why thinking is so hard, but how we can all learn to do it better together.

THE SPEED READ

Why are we so bad at thinking? Partly because it is just too much hassle. It takes time, we are too busy on social media, we don't want to have to change our minds about anything and we are fundamentally lazy. But if we can learn to use System 2, the rational, slow, meditative part of our brains, we can learn to steer our intuitive selves to think better, just as a skilful rider can steer a charging elephant in the right direction.

Wait, What?
And Life's Other Essential Questions

James Ryan, 2017

The original version of this book by James Ryan, dean of education at Harvard University, came when he gave a graduation speech. The hook of the speech was that there are only five questions you really need to know about in life. The speech went viral and inevitably an editor called and asked him to turn it into a book.

That might sound a dispiriting origin story, but Ryan addresses this with charm right from the start. 'Before you roll your eyes, or worse yet, put down this book, let me say this: I completely get that [the idea] might seem grandiose and even a bit outlandish. My only excuse is that this book started as a graduation speech.'

And the book maintains that level of charm throughout what is a genuinely interesting exploration of the theme. Ryan writes, 'Asking good questions is hard because it requires you to see past the easy answers and to focus instead on the difficult, the tricky, the mysterious, the awkward, and sometimes the painful.' He describes questions as the keys that can unlock a door (which you might not even have known was there in the first place, until you asked the question).

He starts with a meditation on the importance of asking the right question, rather than skipping that stage and immediately looking for the right answer. Asking the right questions draws other people into the problem-solving process, but posing a good question isn't as simple as it might seem. He also discusses the way in which we respond to questions, advising that we learn to recognise a good question when we hear one. We need to see the difference between an awkward but decent question, and a question that is hostile or designed to trip you up. The latter is the only truly bad question.

The main five chapters are each based on one simple question, which Ryan suggests will help to unlock problems without antagonising the people they are asked of. The first question, 'Wait, what?' is described as 'The root of all understanding'. The idea is that we sometimes need to take a step back right at the start of any process to assess the situation and avoid hasty, bad responses. It also allows us to take a bit more time to get to understand people and situations before we move on to action.

The second question, 'I wonder . . . ' has two things to achieve: it is the root of all curiosity, any thought process that leads us to new ideas, and it is also a supremely tactful way of asking certain kinds of questions. For instance, if your household needs to save a hundred pounds a week, you could tell your partner what you think they should do. But how much better to ask, 'I wonder if we could find a way to do this . . . ?' as they will then be drawn into wanting to help, rather than being put on the spot.

The third question is, 'Couldn't we at least . . . ?' This is a way of getting yourself unstuck and taking a tentative step in an uncertain direction: the point is that, to be bold or to make changes, you need to start somewhere, as, in the end, our regrets tend to be things we haven't done rather than things we have done.

The fourth question, 'How can I help?' is designed as a way to deal with any issues that come up in a relationship, whether domestic or professional. Rather than giving in to the saviour complex, in which you present yourself as the superior partner who is going to fix something, this question presents you as equal to the other party, and invites them to collaborate constructively.

The fifth question is 'What truly matters?' This is designed to allow you to separate the crucial issues in life from the trivia, and to focus in on the most important stuff.

Finally, there is a 'bonus question': 'And did you get what you wanted from this life, even so?' This is a quote from a poem by Raymond Carver, who was living with cancer when he wrote it. In some respects, it is the most important question of all, as it acknowledges both that life has challenges and that we can rise above them and find 'joy and contentment', and that is ultimately what will matter when we look back in the end.

This is a joy and a surprise of a book. From the seed of an idea that might seem twee or simplistic, it blossoms into a thriving tree, whose fruits are charm, insight and wisdom.

THE SPEED READ

'And there is no greater gift to bestow on children than the gift of curiosity. Effective leaders, even great ones, accept that they don't have all the answers. But they know how to ask the right questions – questions that force others and themselves to move past old and tired answers, questions that open up possibilities that, before the question, went unseen.'

How to be Right . . . in a World Gone Wrong

James O'Brien, 2018

James O'Brien has been a presenter on the flagship BBC news programme *Newsnight* but his most enduring role is as a talk show host on the UK radio station LBC.

The station features strong voices from across the political spectrum and O'Brien is known as one of the most prominent liberal and anti-populist voices on the station. As such, he has been involved in many spats and debates in interviews and online. Whether or not you see his responses as remarkably calm and reasoned, or unbearably smug and elitist probably depends on your viewpoint. Personally, I would tend to the former view.

Talk radio is an inherently confrontational medium: the hosts of the shows more often succeed in maximising listening figures by managing to rile their opponents into phoning up than they do by bringing people together. And O'Brien is certainly confrontational. Even the title of his book seems deliberately designed to provoke his rabid haters. But it is an interesting read, in spite of the occasionally aggressive tone.

O'Brien specialises in asking his callers questions that are designed to bring out their underlying beliefs. Whether it be a caller complaining about hospital waiting times, whose real bugbear is immigration, or people who are furious about 'benefits scroungers' but who are themselves dependent on family credits, or someone objecting to 'political correctness' because, deep down, they are in fact misogynistic, he will often manage, quite gently, to tease out unconscious motivation.

Often O'Brien will focus on quite simple questions: 'Why do you think that?'; 'How does that actually affect you?'; or 'Can you give me a concrete example?' When it comes to an issue such as Brexit or Trump, he will zero in on a question that brings out the confusion of the true believer. Confronted, for instance, with a caller claiming that all UK laws are made by the EU, he will challenge them to name one such law and how it has created a problem in their own life.

It helps that he is not innately unkind to his callers: he is more willing to attack the rich owners of newspapers and TV channels that focus on populist propaganda for the way they have fostered and encouraged such beliefs. His sincere conviction that people expressing toxic beliefs have often been misled and misinformed by politically motivated propaganda makes the conversations less confrontational than they might otherwise become. Of course, this in itself can come across as patronising – as some take it to mean he thinks the callers are stupid enough to have been brainwashed.

There are numerous transcripts of conversations included in the book, covering topics such as Islam, Brexit, issues of sexuality and gender, Trump and more. These can be a bit awkward to read, as the tone of description veers close to the edge of the author boasting of having 'won' his telephone battles, but his commentary mostly manages to stay just the right side of fair-minded and his acidic wit helps to keep the tone light.

One of the things that helps to make O'Brien seem human is the humility he expresses when it comes to some of his own uncertainties. He writes, 'Despite the title of this book, it is refreshing, in an age of increasingly reductionist and binary debate, to recognise the importance of sometimes saying the three most undervalued words in the English language: "I don't know".' He discusses how some of the attitudes he held in the past, especially on gender issues and trans and gay identity, have been successfully challenged either by callers, or by family and friends. By showing that he can, from time to time, admit to being wrong, he helps to further defuse that deliberately provocative title.

On the whole, this is an engaging read, although those who fundamentally disagree with his opinions will probably not enjoy it so much. It isn't so much a guide to being right as a fascinating tour through the many ways in which the mass media encourages easily falsified beliefs in large parts of the population, and a primer on how to stay calm and find rational means to confront such beliefs. If anyone is going to write the ideal etiquette book for such debate, it probably isn't going to be a successful talk show host, but this comes closer than most might manage.

THE SPEED READ

I win a lot of arguments, and here is how I do it (together with some trophy transcripts of said arguments). A lot of people have demented beliefs that they can't actually justify, but this is often because those ideas have been propagated by media moguls with their own agenda, so the best way to confront them is by calmly trying to get them to explain their exact reasons until they are forced to admit I was right all along. (Oh, and I can be wrong sometimes too. Usually not, though . . .)

The Book You Wish Your Parents Had Read (and Your Children Will Be Glad That You Did)

Philippa Perry, 2019

This is supposed to be a 'parenting book for people who don't buy parenting books'. It's included in this collection because in many ways it can be boiled down to advice on how to think about your relationship with your children.

One of the obvious differences between people who are good with children and those who aren't is the level of empathy they show for the infant. When a baby or toddler is crying the main task is often to establish what it is that is actually bothering them. It can be sore gums, a feeling of being left alone for too long, hunger, pain or one of a hundred other things, and it is crucial to really try to understand what your child is trying to tell you, while feeling frustrated you don't understand them.

Philippa Perry is a psychotherapist and the wife of Grayson Perry (as a couple they have two entries: see p.225). Her book doesn't lay down any rules about how to deal with sleep, feeding or any of the usual child-raising tips. Instead it challenges the reader to think about their own relationship with their parents, and to ponder what impact that is now having on their own relationship with their children. The book includes exercises to help you do this: one part of this is to identify your triggers, the things that make you feel anxious or sad, and to avoid being too judgmental about yourself or your parents.

And it is indeed empathy and compassion that Perry focuses on the most. She bases this partly on her personal experience, relating the kinds of 'aha' moments she had when she worked out what a child wanted from her and chose how to respond.

There is also a strong focus on communication. Early in the book she recounts a story in which a woman's daughter got stuck on a climbing frame while the mother was standing nearby checking her phone. The child was quite rattled by the experience and, the following week, showed reluctance to get on the next climbing frame she encountered. This time her mother put the phone away and held her hand as she climbed, and was on hand to deal with any difficulties that arose.

The key thing here was the conversation they had afterwards. The girl asked her mother why she hadn't helped the previous week. After a moment's thought, the mother said that she felt her mother had always wrapped her in cotton wool and not let her make her own mistakes. As a result she actively wanted to allow her kids more freedom so they could learn more self-confidence. The girl's reply was that she had just thought that her mother didn't care.

This is the kind of insight that comes throughout the book. Parents' ideas are rooted in their own experiences but they fail to realise how differently their children are thinking about the same issues until they communicate with them.

The book can be overly simplistic: one of the goals of psychotherapy is to find sources of shame and self-doubt and to help the patient to be compassionate with themselves and to learn how to turn that shame into something closer to pride.

Perry offers some fairly brief strategies for achieving this but she herself will know that in reality this can be a much more complex task to achieve.

The other potential criticism of the book is that it can't offer a solution for all parenting problems: these can arise from so many sources that it is reductive to expect that they can all be solved simply with compassion and empathy. However, so long as you bear that in mind, it is a really interesting read. And, crucially, it can help you to really examine the way you think about parenting in general and, potentially, to realise how important it can be simply to listen to your child and to show them that you have understood what they are saying and regard it as valid.

THE SPEED READ

A parenting guide for people who don't read parenting books: psychotherapist Philippa Perry explores the role of compassion and empathy in parenting, urging the reader to examine their own parental relationships and to compassionately understand how those formative experiences may be affecting their own parenting decisions.

Index